QUANTUM STATISTICS AND THE MANY-BODY PROBLEM

QUANTUM STATISTICS AND THE MANY-BODY PROBLEM

Edited by

Samuel B. Trickey, Wiley P. Kirk,

and

James W. Dufty

University of Florida
Gainesville, Florida

PLENUM PRESS · NEW YORK AND LONDON

Library of Congress Cataloging in Publication Data

Symposium on Quantum Statistics and Many-Body Problems, 1st, Sanibel Island, Fla., 1975.
Quantum statistics and the many-body problem.

Includes bibliographical references.
1. Liquid helium—Congresses. 2. Superfluidity – Congresses. 3. Quantum statistics—Congresses. 4. Problem of many bodies – Congresses. I. Trickey, Samuel B. II. Kirk, Wiley P. III. Dufty, James W. IV. Title.
QC145.45.H4S9 1975 530.1'4 75-25547
ISBN 0-306-30887-8

Proceedings of the first Symposium on Quantum Statistics and Many-Body Problems held on Sanibel Island, January 26—29, 1975

A Division of Plenum Publishing Corporation
227 West 17th Street, New York, N.Y. 10011

United Kingdom edition published by Plenum Press, London
A Division of Plenum Publishing Company, Ltd.
Davis House (4th Floor), 8 Scrubs Lane, Harlesden, London, NW10 6SE, England

Printed in the United States of America

PREFACE

The present volume represents the great majority of the papers presented on Sanibel Island at the first Symposium on Quantum Statistics and Many-Body Problems (January 26-29, 1975). In his Introductory Remarks, Professor Löwdin outlines the history of the original Symposia, and the genesis of the conference whose papers comprise this volume. We join him in his expression of thanks, and note, additionally, our gratitude to him and to Professors N. Y. Öhrn, J. R. Sabin, E. D. Adams, and John Daunt.

The papers are grouped somewhat differently from their order of presentation. It seemed convenient to begin with the six papers which deal with sound propagation in one form or another, then have a two-paper diversion into solid Helium. The ^{3}He superfluid theme is picked up again with four papers on spin dynamics, orbit waves, etc., followed by a selection of five papers on a variety of experimental and theoretical aspects of the ^{3}He superfluid problem. Work in the areas of films, monolayers, and mixtures is presented next, followed by two papers on liquid ^{4}He. We conclude with a selection of six papers on other quantum fluids and general statistical mechanics.

We are most grateful to the contributors to this volume for their patience and cooperation; they have had as editors three utter novices! We have learned much, both scientifically and editorially. We hope that this volume will be of at least some help to others as well.

Samuel B. Trickey
Wiley P. Kirk
James W. Dufty

Gainesville, 10 June 1975

CONTENTS

INTRODUCTORY REMARKS

The Quantum Theory Project at the University of Florida was begun in 1960 as the American part of a new University of Uppsala - University of Florida exchange program in quantum sciences. From 1963 to the present, a series of annual International Symposia on the Quantum Theory of Atoms, Molecules, and Solids has been organized by the Quantum Theory Project and its sister project, the Quantum Chemistry Group, University of Uppsala. These January meetings at Sanibel Island typically attract 250 researchers from all over the world. It has been customary to dedicate the odd-year symposia to one of the outstanding pioneers in the quantum theory of matter: Professors E. A. Hylleraas (1963), R. S. Mulliken (1965), J. C. Slater (1967), H. Eyring (1969), J. H. Van Vleck (1971), E. U. Condon (1973) and L. H. Thomas (1975).

For some years the Sanibel Symposia included sessions devoted to Quantum Statistics and to Quantum Biology. In January, 1974, Quantum Biology became the subject of a three-day Symposium in its own right. The success of this new Symposium prompted Professor L. H. Nosanow (then Chairman of the Department of Physics and Astronomy at University of Florida) to suggest an experiment with Quantum Statistics. Thus, it was decided to arrange a special three-day session on Quantum Statistics and Many-Body Problems as the last of the 1975 series of Symposia. The Quantum Statistics Symposium was from the beginning a joint venture between the Quantum Theory Project and the Department of Physics and Astronomy of the University of Florida.

The Editors of these Proceedings were therefore asked to undertake the detailed organization of the new Symposium, with Professor Trickey as Associate Director. We were fortunate to have Professor David M. Lee of Cornell University as a Visiting Professor at the University of Florida during the academic year 1974-'75, and we are most grateful to Dr. Lee for utilizing so many of his international contacts and putting so much time and effort into the organization of this Symposium. Because of the interests of the organizers, it finally turned out to be appropriate to focus the Symposium on Quantum Fluids, with particular emphasis on superfluid ^{3}He.

On behalf of all the organizers, I take this opportunity to express our gratitude to various units of the University of Florida for financial support of this meeting: the Office of Academic Affairs, the Division of Continuing Education, the Graduate School, the College of Arts and Sciences, and the Department of Physics and Astronomy.

Many of the practical details of the Symposium were handled by faculty, post-doctoral associates, and graduate students of the Quantum Theory Project and of the Department of Physics and Astronomy. Their efforts, as well as those of their fine secretarial staffs, are hereby gratefully acknowledged. We are further indebted to the owners of the Island Beach Club, Mr. Walter P. Condon, Trustee, and the manager, Mr. Robert J. Houser, and their staff for their efforts to provide a pleasant and effective environment for the Symposium on Sanibel Island.

Per-Olov Löwdin

Professor and Head
Department of Quantum Chemistry
University of Uppsala
Uppsala, Sweden

Graduate Research Professor
Quantum Theory Project
University of Florida
Gainesville, Florida

Member of Nobel Committee in Physics

ATTENUATION AND DISPERSION OF FIRST SOUND NEAR THE SUPERFLUID TRANSITION OF LIQUID HELIUM

F. Pobell

Institut für Festkörperforschung, Kernforschungsanlage

517 Jülich, W. Germany

Summary: The superfluid phase transition of liquid helium has been investigated in recent years in great detail and with very high precision. In this paper acoustic investigations of this phase transition will be discussed.

Low frequency measurements of the velocity u of first sound near T_λ[1] have been found to agree with the following Pippard-Buckingham-Fairbank relation[2] as modified by Ahlers[3]

$$u(T)-u(T_\lambda) = A_1/C_p+A_2t+A_3t^{1-\alpha} \quad (1)$$

(C_p: specific heat; $t=|1-T/T_\lambda|$; α: critical exponent of C_p).

Williams and Rudnick[4] have measured the attenuation ($600 \text{ kHz} \leq \omega/2\pi \leq 3.17$ Mhz), and Thomlinson and Pobell[5] have investigated the dispersion ($5.4 \text{ kHz} \leq \omega/2\pi \leq 208$ kHz) of first sound in ^{4}He near T_λ ($1\mu K \leq |T-T_\lambda| \leq 5$ mK). Both are asymmetric around T_λ with a peak on the low temperature side. The measured attenuation and dispersion are discussed as arising from the Landau-Khalatnikov relaxation process occurring only below T_λ, and from a fluctuation process occurring on both sides of the λ-transition. The data can only be compared to the relaxation process because no detailed prediction for the contribution arising from fluctuations is available at present.

Recently these experiments have been extended to ^{3}He-^{4}He mixtures.[6-8] The low-frequency velocity could again be fit to Eq. 1.[6] The dispersion D and attenuation α have been measured in mixtures containing up to 40% ^{3}He and at $2.25 \text{ kHz} \leq \omega/2\pi \leq 594$ kHz.[7,8] Both

D and α behave qualitatively similar as in ^{4}He near T_λ, but their absolute values are considerably reduced when the ^{3}He concentration is increased. The temperature and concentration dependence of the order parameter relaxation time deduced from these measurements[8] is in agreement with results from an independent second sound experiment.[9]

References

(1) M. Barmatz and I. Rudnick, Phys. Rev. 170, 224 (1968)

(2) M. Buckingham and W. Fairbank, Progress in Low Temperature Physics, ed. C. J. Gorter (North Holland, Amsterdam, 1961), Vol. III, Ch. 3.

(3) G. Ahlers, Phys. Rev. 182, 352 (1969).

(4) R. D. Williams and I. Rudnick, Phys. Rev. Lett. 25, 276 (1970).

(5) W. C. Thomlinson and F. Pobell, Phys. Rev. Lett. 31, 283 (1973).

(6) W. C. Thomlinson and F. Pobell, Phys. Lett. 44A, 155 (1973).

(7) C. Buchal, F. Pobell, and W. C. Thomlinson, Phys. Lett. (March, 1975).

(8) C. Buchal and F. Pobell, to be published.

(9) W. C. Thomlinson, G. G. Ihas, and F. Pobell, to be published in Phys. Rev. (1975).

HYDRODYNAMIC THEORY OF FOURTH SOUND IN A MOVING SUPERFLUID - A DISCUSSION OF SIZE EFFECTS

David J. Bergman†

Bell Laboratories

Murray Hill, New Jersey 07974

Abstract: A previous discussion of fourth sound in a moving superfluid is extended to include a detailed estimate of size effects. Comparison with experiment still shows a considerable disagreement. This might indicate the existence of relaxation phenomena on a time scale of 10^{-2} - 10^{2} sec. which have not yet been observed.

A theoretical discussion of fourth sound in an incompressible superleak filled with superfluid ^{4}He in which a static superfluid flow already exists has recently been given by Bergman *et al.*[1], based upon the exact hydrodynamic equations developed for such a system by Halperin and Hohenberg.[2] As a result, one could get an exact equation for the velocity of fourth sound c_4 in the presence of a static superfluid velocity $\tilde{u}_s$

$$c_4 = \left(\frac{\partial\tilde{\mu}}{\partial\tilde{\rho}}\right)_{\tilde{S}\,\tilde{u}_s} \left[\tilde{\rho}_s + \tilde{u}^2_{s||} \left(\frac{\partial\tilde{\rho}_s}{\partial\frac{1}{2}\tilde{u}_s^2}\right)_{\tilde{\rho}\,\tilde{S}} + \tilde{u}_{s||}\left(\frac{\partial\tilde{\rho}_s}{\partial\tilde{\rho}}\right)_{\tilde{S}\,\tilde{u}_s} \right]^{1/2} \tag{1}$$

Here $\tilde{S}$ is the total entropy of the helium and the superleak per unit of the total combined volume, $\tilde{\rho}$ is the helium density per unit of

---

†On leave of absence from Physics Department, Tel-Aviv University, Tel-Aviv, Israel.

the total combined volume, $\tilde{\rho}$ is the helium density per unit of the total volume, $\tilde{\rho}_s$ is the effective superfluid density per unit of the total volume, $\tilde{\mu}$ is the helium chemical potential, and $\tilde{u}_{s||}$ is the component of $\tilde{u}_s$ along the wave vector. Note that $\tilde{u}_s$ is defined as the gradient of the spatially averaged macroscopic phase variable $\tilde{\phi}$

$$\tilde{u}_s \equiv \frac{\hbar}{m} \nabla\tilde{\phi} \tag{2}$$

and is thus in general not equal to the average of any local helium velocity inside the superleak. $\tilde{\mu}$ and $\tilde{\rho}_s$ are defined by suitable derivatives of the total energy per unit of the total volume $\tilde{E}$:

$$\tilde{\mu} \equiv \left(\frac{\partial\tilde{E}}{\partial\tilde{\rho}}\right)_{\tilde{S}\,\tilde{u}_s} \tag{3}$$

$$\tilde{\rho}_s\,\tilde{u}_s \equiv \left(\frac{\partial\tilde{E}}{\partial\tilde{u}_s}\right)_{\tilde{\rho}\,\tilde{S}} \tag{4}$$

Note that, except for the assumption that the superleak is incompressible (this is avoidable) and that there is no migration of vortices in the system, Eq. (1) is an exact consequence of our basic notions concerning superfluid helium in a superleak, i.e., the existence of a complex order parameter. Its experimental verification would be a welcome confirmation of the correctness of these notions, similar to the experimental verification of the second sound equation in bulk helium. For that purpose, one would have to measure independently the thermodynamic quantities $\tilde{\mu}$ and $\tilde{\rho}_s$ at least as functions of $\tilde{\rho}$ and $\tilde{S}$ for $\tilde{u}_s = 0$, and this would have to be done in the same system in which c_4 is measured, since these quantities will vary from superleak to superleak.

Unfortunately, such a detailed investigation has not yet been undertaken for any superleak. On the other hand, some detailed measurements of c_4 in the presence of a static velocity $\tilde{u}_s$ have been performed by Kojima[3] in a superleak where the pores are large enough so that one might expect size effects to be relatively unimportant. In the limit when size effects can be ignored completely, and when the heat capacity of the filler is negligible compared to that of the helium in the superleak, it has been shown in Ref. 1 that Eq. (1) can be reduced to an expression that depends only on thermodynamic properties of pure bulk helium, as well as one geometrical parameter g which varies from superleak to superleak but is independent of the state of the system. Since there appeared to be a rather large disagreement between the theory and the experimental results, we will now discuss the possibility that size effects are important.

However, rather than attempt to give a truly microscopic description of the system in terms of atomic variables, we will assume that the helium inside the pores of the superleak can be described by the local thermodynamic variables of pure bulk helium. In this picture, the size effects are due only to the presence of an external force field coming from the Van-der-Waals forces between the filler atoms and the helium atoms. In that case, most of the thermodynamic functions of helium in the superleak are given by simple averages of local functions of pure helium, such as

$$\tilde{\rho} = \frac{1}{V} \int \rho(r)\, d^3r \quad , \tag{5}$$

where V is the total volume of the helium and the filler, while the integral is only over the helium volume. The superfluid density $\tilde{\rho}_s$ is given by

$$\tilde{\rho}_s = \frac{1}{V} \int \rho_s(r)(\nabla\phi_1)^2 \; d^3r \tag{6}$$

$$\nabla\phi_1 \equiv \frac{v_s(r)}{\tilde{u}_s} \tag{7}$$

where $v_s(r)$ is the local superfluid velocity of the helium inside the pores, and $\rho_s(r)$ is the local superfluid density.

As long as $\tilde{u}_s$ is small and the radii of curvature of the helium-filler interface are much greater than the range of the inhomogeneities in $\rho_s(r)$, $\nabla\phi_1$ is independent of the state of the system: It depends only on the detailed geometry of the superleak and, like g, will vary from superleak to superleak. In that case, the fourth sound velocity can be written as

$$c_4 = c_{40} + \Delta c_4 \tag{8}$$

$$c_{40} \equiv \tilde{\rho}_s \left(\frac{\partial\tilde{\mu}}{\partial\tilde{\rho}}\right)_{\tilde{S}} \Bigg|_{\tilde{u}_s=0} = \frac{\int \rho_s (\nabla\phi_1)^2}{\int \rho \left(\frac{\partial\rho}{\partial P}\right)_T} \frac{1}{1-A^2/CD} \tag{9}$$

$$\Delta c_4 \equiv \tilde{u}_{s_{||}} \left(\frac{\partial\tilde{\rho}_s}{\partial\tilde{\rho}}\right)_{\tilde{S}} \Bigg|_{\tilde{u}_s=0} = \frac{\int \rho \left(\frac{\partial\rho_s}{\partial P}\right)_T (\nabla\phi_1)^2}{\int \rho \left(\frac{\partial\rho}{\partial P}\right)_T} \frac{1}{1-A^2/CD} - \frac{EA}{CD-A^2} \tag{10}$$

where clearly c_{40} is the velocity at $\tilde{u}_s = 0$, while Δc_4 is the Doppler shift, and

$$A \equiv \int \left[\left(\frac{\partial\rho}{\partial T}\right)_P + S \left(\frac{\partial\rho}{\partial P}\right)_T \right] \tag{11a}$$

$$C \equiv \int \rho \left(\frac{\partial\rho}{\partial P}\right)_T \tag{11b}$$

$$D \equiv \int \left[\frac{C_P}{T} + \frac{2S}{\rho} \left(\frac{\partial\rho}{\partial T}\right)_P + \frac{S^2}{\rho} \left(\frac{\partial\rho}{\partial P}\right)_T \right] \tag{11c}$$

$$E \equiv \int \left[\left(\frac{\partial\rho_s}{\partial T}\right)_P + S \left(\frac{\partial\rho_s}{\partial P}\right)_T \right] (\nabla\phi_1)^2 \tag{11d}$$

Numerical estimates indicate that in the superleak FSR IIIa used by Kojima[3] and at $T = 1.2°K$, the size effect corrections only need to be considered in connection with the integrals which remained explicit in Eqs. (9) and (10).

In order to compare our results with this experiment, we focus our attention on the following quantity

$$B \equiv \frac{\Delta c_4 \cdot c_1^2}{\tilde{u}_{s\|} \cdot c_{40}^2} = \frac{\int \rho_s \left(\frac{\partial\rho}{\partial P}\right)_T (\nabla\phi_1)^2 + \int \rho^2 \left(\frac{\partial(\rho_s/\rho)}{\partial P}\right)_T (\nabla\phi_1)^2 - \frac{EA}{D}}{\int \rho_s (\nabla\phi_1)^2 \cdot \left(\frac{\partial\rho}{\partial P}\right)_{S,\text{ bulk helium}}}, \tag{12}$$

where c_1 is the first sound velocity in pure bulk helium. B would be exactly 1 at $T = 0$ if size effects were absent. It also has the advantage that as long as the size effects are not too great, most of the dependence on $\nabla\phi_1$ cancels out between the numerator and the denominator.

Using standard bulk helium data, as well as a Van-der-Waals force of the form α/z^3 (where z is the distance from the helium-filler interface) with $\alpha = 27$ deg·(atomic layers)^3, as determined by Sabisky and Anderson[4], we estimate the three terms in (12) to be

$$.98 - .99; \quad \geq -.09; \quad .04 \tag{13}$$

respectively, which leads to a value $B \geq .93$. The reason for the inequalities is that in the second term of (12), $(\nabla\phi_1)^2$ is weighted more heavily near the helium filler interface, where it is usually smaller than elsewhere because the local flow pattern must follow the contour of the interface. Note also that while the first term of (12) and (13) includes only a temperature independent size correction, and the last term is essentially a size independent temperature correction which vanishes at $T = 0$, the middle term provides corrections of both types.

In the above-quoted experiment[3], both c_{40} and Δc_4 were measured in the same ring shaped superleak at $T = 1.2^\circ K$, in a situation where $\tilde{u}_s$ was presumably known, leading to a value $B = .80$. The discrepancy between our estimate and this experimental value, which seems to be borne out by further, unpublished results of similar experiments[5], could be due to the existence of a relaxation mechanism enabling the static dc flow of the persistent current to occur in a more stable flow pattern than the ac flow of the fourth-sound wave. As discussed in Ref. 1, this decay must take place on a time scale between 10^{-2} and 10^2 sec., and should be observable in a suitably designed experiment. It might also be related to the fact that the dc velocity in these experiments was always much greater than the ac superfluid velocity in the fourth sound wave.[5]

To summarize, we have made detailed estimates of the effects of finite pore size and finite temperature on the velocity of fourth sound and the Doppler shift for ^4He in a superleak. These corrections have reduced the discrepancy previously found between theory and experiments[1], but it still remains fairly large. We would like to point out that our estimate for B is subject to a rather large uncertainty due to the lack of an accurate value for $(\partial u_2/\partial P)_T$ at $P = 0$ (u_2 is the second sound velocity in pure helium), and due to our neglect of other possible size effects which are not included in the local description of the system (e.g., boundary conditions on the magnitude of the complex order parameter). These uncertainties affect mostly the second term of (12), which also gives the largest contribution to 1-B. Consequently, it cannot be ruled out completely that size and temperature effects may be able to account for the discrepancy in B. All we can say is that our present best estimate is still about 15% higher than the experimental value.

ACKNOWLEDGMENTS

B. I. Halperin and P. C. Hohenberg collaborated with me on the earlier parts of this work, which were reported in Ref. 1. I would also like to acknowledge useful conversations with C. Herring, H. Kojima, J. D. Reppy, and I. Rudnick, and to thank B. I. Halperin for a critical reading of the manuscript.

References

(1) D. J. Bergman, B. I. Halperin, and P. C. Hohenberg, submitted to Phys. Rev. B.

(2) B. I. Halperin and P. C. Hohenberg, Phys. Rev. 188, 898 (1969). The application of these results to helium in pores was spelled out in greater detail by P. C. Hohenberg in Physics of Quantum Fluids (1970) Tokyo Summer Lectures in Theoretical Physics) edited by R. Kubo and T. Takano, (Syokabo, 1971).

(3) H. Kojima, Ph.D. Thesis, UCLA. See also H. Kojima, W. Veith, S. J. Putterman, E. Guyon, and I. Rudnick, Phys. Rev. Lett. 27, 714 (1971); and H. Kojima, W. Veith, E. Guyon, and I. Rudnick, J. Low Temp. Phys. 8, 187 (1972).

(4) E. S. Sabisky and C. H. Anderson, Phys. Rev. Lett. 30, 1122 (1973); Phys. Rev. A7, 790 (1973).

(5) I. Rudnick, private communication.

THEORY OF SOUND ABSORPTION IN THE SUPERFLUID PHASES OF ^{3}He

P. Wölfle

Max-Planck-Institut für Physik und Astrophysik

D-8000 München 40, Germany

After the discovery of the superfluid phases of ^{3}He,[1] the propagation of zero sound was among the first properties to be studied experimentally.[2,3] Somewhat unexpectedly, the sound attenuation at frequencies exceeding about 5 MHz was observed to develop a rather narrow peak just below the normal-superfluid transition, growing rapidly as the frequency was increased. It was soon realized by several authors[4,5,6] that the sound absorption peak could be understood in the framework of BCS-theory (as applied[7] to the new phases of ^{3}He) as being caused by excitations of the pair condensate. There are two types of absorption processes: (1) breaking of Cooper pairs followed by excitation of quasiparticle pairs of minimum energy $2\Delta(\hat{k})$, where $\Delta(\hat{k})$ is the energy gap along the direction $\hat{k}$ in momentum space, and (2) excitation of collective modes of the order parameter of typical frequency $\omega \sim \Delta(T)$.[8] While the former processes also occur in superconductors (s-wave pairing), the latter processes are characteristic of $L \neq 0$ pairing and may be imagined as vibrations of the internal structure of the Cooper pairs.

Given this interpretation, it is obvious that experimental studies of sound propagation are a useful tool for probing the nature of correlations, or more specifically the symmetry of the pair order parameter, in the low temperature phases of liquid ^{3}He, provided a quantitative theory of these phenomena is available. Such a theory on the level of Landau's Fermi liquid theory has been given in two essentially equivalent forms, first on the basis of a generalized kinetic theory of Fermi liquids[4] and somewhat later in the formulation of a generalized RPA theory.[6]

Detailed calculations[4] have been performed for the axial p-wave state, characterized by the gap parameter matrix $\Delta_{k\alpha\beta}(T) = \Delta_o(T)[(\hat{n}_1\cdot\hat{k})+i(\hat{n}_2\cdot\hat{k})]\hat{V}\cdot(\vec{\sigma}i\sigma^y)_{\alpha\beta}$, where $\hat{n}_1$, $\hat{n}_2$ $((\hat{n}_1\cdot\hat{n}_2)=0)$ and $\hat{V}$ are unit vectors constituting the structure of the order parameter. In this case there are two collective modes, a "clapping" mode[8a] at frequency $\omega_\alpha = 2[{}^1/_5(2\sqrt{6}-3)]^{1/2}\Delta_o(T) \simeq 1.23\,\Delta_o(T)$ and a "flapping" mode at $\omega_\beta \cong (4/5)^{1/2}\Delta_o(T)$, which may be visualized as in Fig. 1. These modes are damped by pair-breaking processes. The sound attenuation was found to be strongly anisotropic, mainly because the coupling of the collective modes to the sound is very sensitive to the angle formed by the axis of the gap $\hat{\ell}$ and the direction of sound propagation $\hat{q}$. For $\hat{q}\,||\,\hat{\ell}$, the pair oscillation modes are decoupled and a relatively broad peak originating from pair-breaking processes is found. For $\hat{q} \perp \hat{\ell}$, only the "clapping" mode couples, resulting in a very sharp attenuation peak. For all intermediate orientations both modes couple, the "flapping" mode producing a second, much broader peak at a somewhat lower temperature. Any averaging over orientations essentially produces a two-peak structure.[8b] Comparing the theory with data[2,3] in the A-phase it appears that while temperature scale and size of the attenutaion, as well as frequency dependence and pressure dependence are roughly described correctly, there remain significant discrepancies. The experimentally observed peak is broader (a second peak is not

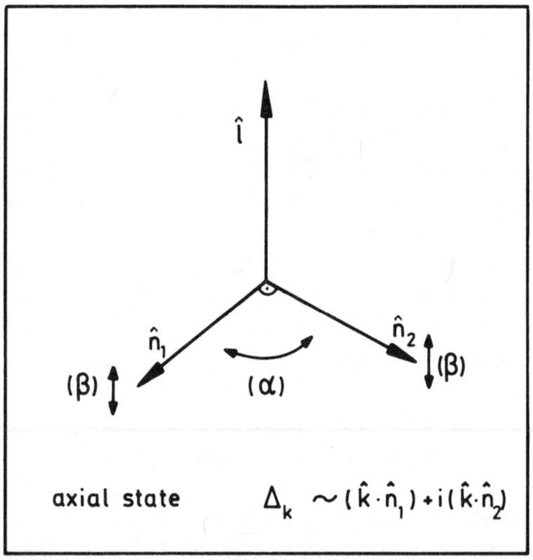

Figure 1. Geometrical representation of the two collective modes in the axial p-wave state. Mode (α) consists of in plane oscillations of $\hat{n}_1$ and $\hat{n}_2$ against each other, mode (β) of in phase oscillations of $\hat{n}_1$ and $\hat{n}_2$ out of plane. The significance of the names "clapping mode" for (α) and "flapping mode" for (β) is self-evident.

resolved) and the anisotropy is less pronounced.[9,10]

The theory has also been worked out for the quasi-isotropic (Balian-Werthamer) state with gap parameter $\Delta_{k\alpha\beta}=\Delta(T)(R\hat{k})(\vec{\sigma}i\sigma^y)_{\alpha\beta}$ (R is a rotation matrix), which is tentatively identified with the B-phase. There a single collective mode at $\omega_c=(12/5)^{1/2}\Delta(T)$ is found,[11] corresponding to an oscillation of the orbital structure (a geometrical picture is difficult to imagine in this case). This mode gives rise to a delta-function peak in the sound attenuation at a temperature somewhat lower than the absorption band caused by pair-breaking (the collective mode is not damped by pair-breaking because $\omega_c < 2\Delta$ and $\Delta(T)$ is isotropic). Experimentally a single attenuation peak somewhat narrower than the one in the A-phase has been observed.[3,16]

Barring possibly different pairing symmetries (e.g., f-wave pairing), the disagreement of theory and experiment suggests that an important element is missing in the theories described so far. It is the main result of the present paper to show that this element is given by the effect of collisions on the sound propagation, especially on the damping of collective modes. The previous theories were aimed at describing the equivalent of zero sound (or "collisionless" sound) in the pair-correlated phases. There it was plausibly assumed that collision effects could be separated from pair excitation effects, both contributing additively to the sound absorption. This assumption was based on the fact that for typical values of sound frequency (~10 MHz), temperature (~2.5 mK) and pressure (~20-34 bar) the collision-induced width of the sound mode is roughly $(\Delta\omega/\omega)\sim 10^{-3}$ (the pair correlation induced width at peak maximum for 20 MHz sound is $(\Delta\omega/\omega)\sim 10^{-2}$) and thus represents a very small perturbation. However, the latter figure does not give a fair account of the importance of collisions in ^{3}He at the transitions, because the effectiveness of collisions in destroying the coherence of the sound wave is strongly inhibited by conservation laws. A more significant quantity is the quasiparticle lifetime τ, which, although not measurable directly, may be inferred from the sound attenuation in the normal Fermi liquid.[12] For the parameters of interest to us here one obtains $\omega\tau\sim 2-8$. It is clear that a broadening of the collective modes of order $\frac{1}{\tau}$ would give rise to substantial changes in the theoretically predicted sound attenuation, in particular for the quasi-isotropic state.

In the following, a theory of sound propagation in the superfluid phases of ^{3}He which incorporates the effect of quasiparticle collisions in a consistent and realistic fashion, is described briefly. Results are reported for the quasi-isotropic state, where collision effects are thought to be most important and the calculations are somewhat less tedious due to the isotropy of the energy gap. In the kinetic equation formulation of the theory,[4] collision effects may be taken into account in a natural way by adding a collision integral to the collisionless kinetic equation given earlier[4] (for notation and definitions we refer to Ref. 4). The full kinetic equation, valid for frequencies $\omega \ll e_F$ and wavevectors $q \ll k_F$ (e_F and k_F are Fermi energy and wavevector, respectively) reads[13]

$$\underline{\Omega}\{\delta\underline{n}\} - \delta\underline{Q} = -i\ \underline{I}\{\delta\underline{n}'\} \tag{1}$$

where

$$\underline{\Omega}\{\delta\underline{n}\} = \omega\delta\underline{n} - \delta\underline{n}\ \underline{e}^0_{\vec{k}_+} + \underline{e}^0_{\vec{k}_-}\ \delta\underline{n}$$

and

$$\delta\underline{Q} = \underline{n}^0_{\vec{k}_-}\ \delta\underline{\varepsilon} - \delta\underline{\varepsilon}\ \underline{n}^0_{\vec{k}_+}$$

Here $\delta\underline{n}_{\vec{k}}(q,\omega)$ is the matrix distribution function (underlined symbols denote 2x2 matrices in particle-hole space), $\delta\underline{\varepsilon}_{\vec{k}}(q,\omega)$ is the matrix generalization of Landau's quasiparticle energy and $\underline{n}^0_{\vec{k}}$ and $\underline{\varepsilon}^0_{\vec{k}}$ are the respective equilibrium quantities, $\vec{k}_{+,-} = \vec{k} \pm \vec{q}/2$. The collision integral has been derived from the microscopic theory.[13] The microscopic expression for the matrix elements of the collision integral has the form

$$I_{k_1\mu\mu'} = (1/12) \sum_{k_2,k_3,k_4} \sum_{\nu_1,\ldots\nu_4} \sum_{\mu_1,\ldots\mu_4',\bar{\mu}_1,\bar{\mu}_1'}$$

$$\times\ T(\mu_1k_1,\mu_2k_2;\mu_3k_3,\mu_4k_4)T^*(\mu_1'k_1,\ldots\mu_4'k_4)(2\pi)^4$$

$$\times\ \delta(\vec{k}_1+\vec{k}_2-\vec{k}_3-\vec{k}_4)\delta(\nu_1E_1+\nu_2E_2-\nu_3E_3-\nu_4E_4) \tag{1a}$$

$$\times \left[\prod_{i=1}^{4} 2\cosh(E_i/2T)\right]^{-1} a_{\bar{\mu}_1\bar{\mu}_1'}(\nu_1k_1)a_{\mu_2'\mu_2}(\nu_2k_2)a_{\mu_3\mu_3'}(\nu_3k_3)$$

$$\times\ a_{\mu_4\mu_4'}(\nu_4k_4)[\delta_{\mu\mu_1}\delta_{\mu_1'\bar{\mu}_1}\delta_{\mu'\bar{\mu}_1'} + \delta_{\mu_1'\mu'}\delta_{\mu\bar{\mu}_1}\delta_{\mu_1\bar{\mu}_1'}]$$

$$\times\ [\delta\phi_{\bar{\mu}_1\bar{\mu}_1'}(\nu_1k_1)+\delta\phi_{\mu_2'\mu_2}(\nu_2k_2)-\delta\phi_{\mu_3\mu_3'}(\nu_3k_3)-\delta\phi_{\mu_4\mu_4'}(\nu_4k_4)]$$

where μ_i, $\nu_i = \pm 1$ are particle-hole indices (spin indices have been suppressed for simplicity) and

$$\delta\phi_{\mu\mu'}(\nu k) = \frac{1}{2}\frac{1}{f(E)f(-E)}\{\delta n_{k\mu\mu'} - \delta n^{\ell}_{k\mu\mu'} + \nu[\delta r + \frac{df}{dE}\delta r']_{k\mu\mu'}\}/a_{\mu\mu'}(\nu k)$$

($f(E)$ is the Fermi function). The quantities δr and $\delta r'$ vanish for $T \to T_c$ and will be neglected in the following. The one particle spectral function is defined by

$$\underline{a}(\nu k) = \frac{1}{2}[\underline{1} + \nu \underline{e}^0_k / E_k] \quad .$$

The scattering amplitude $T(\mu_1 k_1, ..\mu_4 k_4) = \delta_{\mu_1+\mu_2,\mu_3+\mu_4} t(\mu_1 k_1, ...)$ is assumed to be given by normal Fermi liquid theory up to terms of order (T_c/ε_F).

The collision integral operates on $\delta\underline{n}' = \delta\underline{n} - \delta\underline{n}^{\ell}$, the deviation from the local equilibrium distribution function $\delta\underline{n}^{\ell}$, which obeys the collisionless equation

$$\underline{\Omega}\{\delta\underline{n}^{\ell}\} = \delta\underline{Q} + \omega(df/dE)\,\delta\underline{\varepsilon} \quad . \qquad (2)$$

It is readily seen that in the static limit $\delta\underline{n} = \delta\underline{n}^{\ell}$ and thus the collision integral vanishes, as required by thermodynamic stability.

The sound attenuation and dispersion is obtained from the corresponding eigenmode of Eqs. (1) and (2). It is obvious that an exact solution of this system of integral equations is out of the question. Rather than seeking an approximate solution of the exact equations, we concentrate in the following on the exact solution to an approximate form of Eq. (1), in which the collision integral is replaced by an appropriate relaxation time expression. This approach has the additional benefit of being valid for arbitrary $\omega\tau$.

As is familiar from transport theory, the collision integral (1a) consists of two parts, the direct (or out-) scattering term (operating on $\delta\phi_k$) and the back (or in-) scattering terms. In the vicinity of T_c, [1] it is a good approximation to neglect the off-diagonal elements of the collision operator, such that the direct term is given by $\underline{I}^d_k = \frac{1}{\tau_k}\delta\underline{n}'_k$, where τ_k is the lifetime of quasiparticles of momentum k (the microscopic expression for τ_k is easily derived from Eq. (1a)). The backscattering terms are of the form $\underline{I}^b_{k\mu\mu'} = \Sigma_{k'} K_{kk'\mu\mu'}\delta n'_{k'\mu\mu'}$, where the diagonal ($\mu=\mu'$) elements of the kernel K are different from the off-diagonal ones. The energy dependence of K is approximately given by df/dE near T_c and Δ^2/E^3 for $T \ll T_c$ and may be simulated by the function $d\phi/d\varepsilon$, where $\phi(\varepsilon) = (\varepsilon/2E)\tanh(E/2T)$. In contrast to the normal Fermi liquid, the energy dependence of the distribution function in the correlated phases is not simply given by $df/d\varepsilon$, but has a more complicated and q-dependent structure, as can be seen from the solution to the collisionless equation given in Ref. 4. Since for $L\neq 0$ pairing one has lost rotational invariance, the kernel K depends separately on $\hat{k}$ and $\hat{k}'$. However, in the vicinity of T_c (and a fortiori in the quasi-isotropic state) the dependence on $(\hat{k}\cdot\hat{k}')$ is most important and one may neglect the remaining angular dependences. It is well known that any collision integral has to satisfy certain relations which follow from the existence of invariant parameters in collisions. Here, the conserved quantities are particle number (mass and spin), momentum and energy; consequently, $\Sigma_{k\mu} I_{k\mu\mu} = \Sigma_{k\mu} \vec{k} I_{k\mu\mu} = \Sigma_{k\mu\mu'} \varepsilon^0_{k\mu\mu'} I_{k\mu'\mu} = 0$, as is readily seen from the explicit expression for I, Eq. (1a). These exact relations have to be satisfied by any approximation to the collision integral and in fact put severe constraints on the L=0 and L=1 angular momentum components of the backscattering integral (in the normal Fermi liquid, the L=0 and L=1 components of I vanish).

On the basis of the foregoing discussion, the desired relaxation time model of the collision integral is written as an expansion in terms of Legendre polynomials in $(\hat{k}\cdot\hat{k}')$

$$I_{k\mu\mu} = \frac{1}{\tau}\delta n'_{k\mu\mu} - \frac{d\phi}{d\varepsilon} N_F^{-1} \sum_{k'} \left\{ \frac{1}{\tau}[1+3(\hat{k}\cdot\hat{k}')] + \frac{1}{\tau'_2} 5P_2(\hat{k}\cdot\hat{k}') \right\} \delta n'_{k'\mu\mu} \tag{3a}$$

$$I_{k\mu-\mu} = \frac{1}{\tau}\delta n'_{k\mu-\mu} + \frac{d\phi}{d\varepsilon} N_F^{-1} \sum_{k'} \frac{1}{\tau'_a} \delta n'_{k'\mu-\mu} \tag{3b}$$

where N_F is the density of states at the Fermi level. While the first two terms in the curly brackets of Eq. (3a) ensure particle number and momentum conservation (note $\int d\varepsilon\, d\phi/d\varepsilon = 1$), the last term describes the L=2 component of the diagonal backscattering integral. A term taking care of energy conservation has been dropped; it has very little influence on the sound properties because of the weak coupling of density and energy fluctuations in ^{3}He. It turns out that in applications of the theory to ^{3}He it is not necessary to include higher relaxation terms (L>2), since those give rise to corrections of relative order s^{-2}, where $s=c_o/v_F$ is the ratio of sound velocity to Fermi velocity which varies from ~ 3.5 to ~13 as one goes from low to high pressure (a Legendre function P_L has to be matched by a factor $(\vec{k}\cdot\vec{q})^L \sim s^{-L}$). Similarly, higher terms in Eq. (3b) may be neglected. The relaxation rate $\frac{1}{\tau'_2}$ has been calculated[14] for the normal Fermi liquid and found to be given by $\frac{1}{\tau'_2} = (1=\xi_2)\frac{1}{\tau}$, where $\xi_2 \cong 0.35$, apparently almost independent of pressure. ξ_2 is defined as the ratio of different angular averages of the scattering cross section, and is likely to change little, at least in the vicinity of T_c. The "relaxation rate" $\frac{1}{\tau'_a}$ in Eq. (3b) is derived from the kernel $K_{\mu-\mu}$ and has no equivalent in normal Fermi liquid theory. An estimate of $K_{\mu-\mu}$, assuming isotropic scattering, yields $\frac{1}{\tau'_a} \approx \frac{1}{3}\frac{1}{\tau}$; due to interference effects, the scattering rate is enhanced by the back scattering term, rather than reduced as usual. In the actual calculation, the relatively unimportant $\frac{1}{\tau'}$ -Term was lumped into a $\frac{1}{\tau_a} \cong \frac{4}{3}\frac{1}{\tau}$ in place of $\frac{1}{\tau}$ on the r.h.s. of (3b).

Eqs. (1) and (2), supplemented by the collision integral (3), have been solved[15] for the quasi-isotropic state following the method employed in Ref. 4. The rather lengthy result for the sound dispersion relation may be written in abbreviated form as

$$\frac{s_1}{s_0} = -(4/15s_0^2) \quad \{(1+F_1/3)\frac{i}{2\tilde{\omega}\tau}\,\phi + \xi_0 + \xi_1 F_1/3\} \quad (4)$$

where $s=\omega/v_F q=s_0+s_1$, with $s_0^2=(F_0/3+3/5)(F_1/3+1)$, $\tilde{\omega}=\omega+i/\tau$, and $\phi(T)= -\int d\varepsilon\ df/dE$. The sound attenuation α and velocity shift are expressed in terms of s_1 as $\alpha=-q\mathrm{Im}(s_1/s_0)$ and $(c-c_0)/c_0=\mathrm{Re}(s_1/s_0)$, respectively. In the limit $\Delta \to 0$ one obtains $\xi_0= -\frac{1}{2}(\omega/\tilde{\omega})^2(\omega\tau_2'-i)^{-1}$ and $\xi_1 = -\frac{1}{2}(\omega/\tilde{\omega})(\omega\tau_2'-i)^{-1}(1+\frac{i}{\tilde{\omega}\tau}\frac{3}{F_1})$; substituted in (4) one finds $(s_1/s_0^2)= -i(2/15s_0^2)(1+F_1/3)(\omega\tau_2+i)^{-1}$, in agreement with Pethick's result[12] $(\frac{1}{\tau_2}=\frac{1}{\tau}-\frac{1}{\tau_2'})$. In the collisionless limit, $\xi_0 = \xi_1 = \lambda(T)(\omega^2-4\Delta^2)/(5\omega^2-12\Delta^2)$, where $\lambda(T) = \int d\varepsilon \frac{\Delta^2}{E}\frac{\tanh(E/2T)}{4E^2-\omega^2}$ in agreement with the result found by Maki.[11] Taking collisions into account, ξ_0 and ξ_1 contain broadened collective mode contributions approximately proportional to $[(\omega+i/\tau)(\omega+i/\tau_a)-12\Delta^2/5]^{-1}$, giving rise to a peak in the sound attenuation of width at half maximum $\Delta T_{1/2} = 2[(\omega\tau)^{-1}+(\omega\tau_a)^{-1}](T_c-T_\omega)$, where T_ω is defined by $\omega = (12/5)^{1/2}\Delta(T_\omega)$. The low temperature tail of the attenuation is governed by subtle cancellations of various terms in ξ_0 and ξ_1.

In Fig. 2 the theoretical curves are plotted together with data for the B-phase.[3] The agreement of theory and experiment is good; in particular, the width of the attenuation peak (or, complementary, the peak maximum) is given correctly. In this calculation the following parameter values (appropriate for 19.6 bar pressure) have been used: $T_c=2.35$ mK, $\Delta^2(T)=3.1\ T_c(T_c-T)$ (from $\Delta^2(T)=\frac{2}{3}(\pi T_c)^2(\Delta C/C_N)(1-T/T_c)$, substituting a value for the specific heat discontinuity of $(\Delta C/C_N)=1.5$[3b], $c_0=357$ m/sec, $s_0^2=98$, $F_1=12.2$, $\tau_2(T_c)=1.25\ 10^{-}$ sec (from $\alpha(T_c)=1.54\ \mathrm{cm}^{-1}$). The energy and temperature dependence of $\frac{1}{\tau}$ was roughly estimated from the microscopic expression to be given by $\tau^{-1}(\varepsilon,T)=\tau^{-1}(0,T_c)(T/T_c)^2[1+(\varepsilon/\pi T)^2-8(\Delta/\pi T)^2]$ near T_c. The same dependence on ε and T was assumed for $\tau_2'^{-1}$ and τ_a^{-1}. The theory is also in good agreement with recent data by Roach et al.,[16]

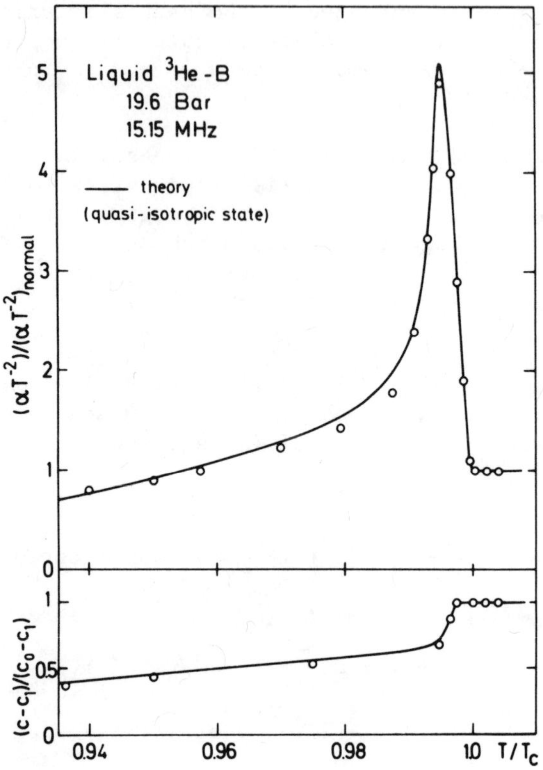

Figure 2. Normalized sound attenuation and velocity shift in ^{3}He-B versus reduced temperature (c_0 and c_1 are the velocities of zero and first sound, respectively). Solid curves: theoretical result for the quasi-isotropic (Balian-Werthamer) state. Circles: representative data points by Paulson et al.[3]

in particular with the velocity data.

Similar calculations for the axial state are in progress. In this case, as mentioned before, the collective modes are already damped by pair breaking. The additional broadening caused by collisions may be estimated to be $\Delta T_{\alpha,\beta} \approx 4(\omega\tau)^{-1}(T_c - T_{\alpha,\beta})$, where $T_{\alpha,\beta}$ is defined by $\omega \approx 1.23\Delta(T_\alpha)$ and $\omega \approx 0.9\Delta(T_\beta)$, respectively. For 20 MHz sound at the melting pressure $\omega\tau \sim 4$; the collision broadening is enough to wipe out the two peak structure! This is true down to pressures somewhat below the polycritical point, in agreement with experiment.

The success of the present model calculation lends support to the identification of the B-phase as the quasi-isotropic (BW) state, which has been promoted by Brinkman and Osheroff on the basis of NMR data.[17] Although the present agreement of theory and experiment does not preclude similar agreement for some f-wave state (yet to be found!), there are two general points one can make here: (1) It is fairly obvious that pairing states of higher angular momentum L will possess an increasing number of collective modes due to the richer structure of the gap parameter (this is supported by work on d-wave pairing).[18] Accordingly one would expect a relatively broadly structured sound attenuation, in contrast to the narrow peak that is actually observed in the B-phase. (2) In order to decide on this question there is still the possibility of reducing the effect of collisions by following the phase boundary to lower pressures and hence lower temperatures. At the saturated vapor pressure the transition temperature is ~0.9 mK[19] and the collision rate, being proportional to T^2, is down by a factor of ~10 from its value at the melting pressure, i.e., the attenuation peak should be 10 times as sharp, if the B-phase is the quasi-isotropic state. If not, one would expect to see some structure, which would give clues as to the type of f-wave pairing involved (for example).

The same possibility is open for the A-phase in a small magnetic field, large enough to generate a small strip of A-phase down to zero pressure, but small enough to avoid the complications associated with the splitting of the A-transition (for which the theory has not been worked out yet). If the A-phase is indeed the axial state, one would expect to see the full two peak structure in the sound attenuation (as discussed in the first section). The observation of the then large anisotropy effects would provide unambiguous evidence for (or against) the axial state.

In conclusion I should like to thank the organizers of the Sanibel Symposium on Quantum Statistics and Many-Body Problems, 1975, for a most effective and pleasant conference as well as financial support.

References

(1) D. D. Osheroff, R. C. Richardson, and D. M. Lee, Phys. Rev. Lett. 28, 885 (1972); D. D. Osheroff, W. J. Gully, R. C. Richardson and D. M. Lee, Phys. Rev. Lett. 29, 920 (1972).

(2) D. T. Lawson, W. J. Gully, S. Goldstein, R. C. Richardson, and D. M. Lee, (a) Phys. Rev. Lett. 30, 541 (1973); and (b) J. Low Temp. Phys. 15, 169 (1974).

(3) D. N. Paulson, R. T. Johnson, and J. C. Wheatley, Phys. Rev. Lett. 30, 829 (1973); J. C. Wheatley, Physica 69, 218 (1973).

(4) P. Wölfle, Phys. Rev. Lett. 30, 1169 (1973); J. W. Serene, Ph.D. thesis, Cornell University 1974 (unpublished).

(5) B. R. Patton, UCSD preprint 1973.

(6) H. Ebisawa and K. Maki, Progr. Theor. Phys. 51, 337 (1974).

(7) P. W. Anderson and W. F. Brinkman, Phys. Rev. Lett. 30, 1108 (1973); R. Balian and N. R. Werthamer, Phys. Rev. 131, 1153 (1963).

(8) Units are chosen such that $\hbar = k_B = 1$.

(8a) These names were suggested to me by W. M. Saslow, while watching sea gulls on the beach.

(8b) The so-called "pair breaking cusp" at T_ω ($\omega=2\Delta(T_\omega)$), sometimes referred to in the literature, is an artifact of an imprecise numerical evaluation of the expression for the sound attenuation. α is a smooth function of temperature at T_ω (J. W. Serene, private communication).

(9) D. T. Lawson, H. M. Bozler, and D. M. Lee, Phys. Rev. Lett. 34, 121 (1975), and these proceedings.

(10) Pat R. Roach, B. M. Abraham, P. D. Roach, and J. B. Ketterson, to be published, and these proceedings.

(11) K. Maki, J. Low Temp. Phys. 16, 465 (1974).

(12) C. J. Pethick, Phys. Rev. 185, 384 (1969).

(13) P. Wölfle, to be published.

(14) K. S. Dy and C. J. Pethick, Phys. Rev. 185, 373 (1969); and unpublished results quoted in Ref. 2(b), Table II.

(15) P. Wölfle, to be published.

(16) Pat R. Roach, B. M. Abraham, M. Kuchnir and J. B. Ketterson, to be published and these proceedings.

(17) D. D. Osheroff and W. F. Brinkman, Phys. Rev. Lett. 32, 584 (1974); D. D. Osheroff, Phys. Rev. Lett. 33, 1009 (1974).

(18) R. Lautkaski and P. Wölfle, unpublished.

(19) A. I. Ahonen, M. T. Haikala, M. Krusius, and O. V. Lounasmaa, Phys. Rev. Lett. 33, 629 (1974).

SOUND PROPAGATION AND ANISOTROPY IN LIQUID ^{3}He A*

D. T. Lawson,[†§] H. M. Bozler,[§] and D. M. Lee[‡§]

[†]Department of Physics, Duke University
Durham, North Carolina 27706

[§]Laboratory of Atomic and Solid State Physics
Cornell University, Ithaca, New York 14850

[‡]Department of Physics and Astronomy
University of Florida, Gainesville, Florida 32611

Abstract: Various orientations of the energy gap of superfluid ^{3}He A may be examined by varying the angle between the applied magnetic field and the direction of sound propagation. Strong evidence of anisotropy has been obtained from both velocity and attenuation measurements.

Nuclear magnetic resonance experiments,[1] in conjunction with the theory of Leggett,[2] led to the hypothesis that liquid ^{3}He A is an anisotropic superfluid in the so-called axial or Anderson-Brinkman-Morel (ABM) state[3,4] of the $\ell = 1$ pairing manifold. This state contains only pairs parallel ($S_z = +1$) and anti-parallel ($S_z = -1$) to the applied magnetic field and is characterized by an anisotropic energy gap with two nodes at opposite poles. Thus the energy gap has an axial symmetry. The relative orbital angular momentum vectors of the various Cooper pairs tend to align in a common direction in this state so that it is possible to define a

---

*Work supported by the National Science Foundation through grants No. GH-35692 and No. GH-38545 and under Grant No. GH-33637 to the Cornell Materials Science Center.
[†]Present address.
[‡]John Simon Guggenheim Memorial Fellow on leave from Cornell University, Ithaca, New York 14850, September 1974 - June 1975.

macroscopic vector $\vec{\ell}$ for each point in the fluid. This vector lies along the direction of the gap axis. The dipolar interaction makes it energetically favorable for the $\vec{\ell}$ vector to be aligned perpendicular to any applied magnetic field. P. G. de Gennes and his collaborators[5,6,7] have shown that the $\vec{\ell}$ vector must be perpendicular to container walls or other boundaries. The combined influences of boundaries, magnetic field, flow, etc., can cause the $\vec{\ell}$ vectors to form patterns called "textures."

In order to explore further the properties of ^{3}He A as well as to test the hypothesis that this phase corresponds to the axial state, we undertook a series of compressional sound propagation experiments in which the angle between the direction of the applied magnetic field and the direction of sound propagation could be varied. A preliminary account of this work recently has been published.[8] The experiment was performed in a Pomeranchuk cell so that the observations were restricted to the ^{3}He melting pressure. Temperatures were obtained from the melting pressure using the data of Halperin et al.[9] Pulse time of flight and attenuation measurements were made using 20 MHz X-cut quartz transducers separated by a 4.14 mm horizontal sound path through the liquid. An external electromagnet could be rotated about the vertical axis of the cell to provide a magnetic field at any orientation in the horizontal plane. Some of the details of the apparatus and the procedure used for the attenuation measurements are described in reference 8; further details will appear in a subsequent publication.

In Fig. 1a the effect of applying a magnetic field to bulk ^{3}He A is shown schematically. The $\vec{\ell}$ vector at a point in the liquid

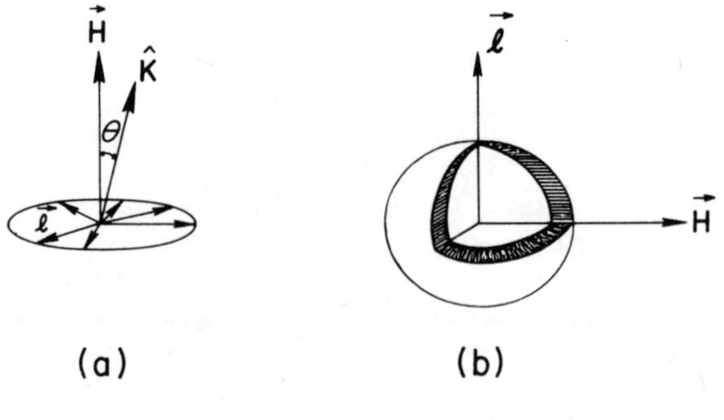

Figure 1. (a) The pair orbital angular momentum vector $\vec{\ell}$ tends to lie in a plane perpendicular to an applied magnetic field. (b) The anisotropic energy gap is represented by a shaded region between the two spheroids. When $\hat{\kappa} \| \vec{H}$, the sound propagation will be perpendicular to the gap axis and the direction of $\vec{\ell}$ in the plane perpendicular to $\vec{H}$ will not influence the attenuation of compressional zero sound. When $\hat{\kappa} \perp \vec{H}$, then $\vec{\ell}$ or the gap axis can be at any angle relative to the $\hat{\kappa}$ vector.

will tilt into the equatorial plane perpendicular to the magnetic field. The direction of the $\vec{\ell}$ vector in the equatorial plane is expected to be random in the absence of external influences, but it should be kept in mind that $\vec{\ell}$ can be oriented by boundaries or flow, including supercurrents induced by thermal gradients.

If we let θ be the angle between the applied field $\vec{H}$ and the propagation direction $\hat{\kappa}$, then for $\theta = 0$, the measured attenuation and velocity of sound will be independent of the distribution of $\vec{\ell}$ in the equatorial plane. The reason for this can be seen in Fig. 1b where the energy gap dependence on angle is shown schematically. As long as $\hat{\kappa} || \vec{H}$, ($\theta = 0$), the direction of sound propagation will correspond to the maximum gap parameter, no matter what direction $\vec{\ell}$ assumes in the plane perpendicular to $\vec{H}$. The situation is quite different for $\theta = 90°$. The gap axis or $\vec{\ell}$ vector then may have any orientation relative to the sound propagation direction. If $\vec{\ell}$ is distributed randomly throughout the liquid sample, then an unweighted average of the attenuation over all angles between $\hat{\kappa}$ and $\vec{\ell}$ should correspond to the experimentally observed attenuation. Figure 2 compares low field attenuation data near the A transition taken at $\theta = 0$, 54°, and 90° with attenuations obtained from the current theory of the axial state,[10] using the assumption that the pair $\vec{\ell}$ vectors are equally distributed among all orientations in the plane perpendicular to the applied field. The high temperature cusp in the theoretical curves corresponds to the breaking of Cooper pairs while the low temperature maxima are associated with collective oscillations have a characteristic spatial variation which is less than the coherence length.[11] While the detailed structure predicted by theory is not evident in our data, the fact that the $\theta = 0$ peak is higher than the $\theta = 90°$ peak whereas in the lower temperature "shoulder" region the $\theta = 0$ attenuation is smaller than the $\theta = 90°$ attenuation is at least consistent with theory. The small amount of spread in the data is a possible indication that the assumption of randomly distributed directions of the $\vec{\ell}$ vector is a reasonable one for the lower fields. Attenuation data obtained in the A phase by Ketterson and co-workers[12] for pressures below the melting pressure show behavior quite similar to that observed in the present experiment.

In a magnetic field the A transition splits into two transitions, A_1 and A_2, with the magnitude of the temperature splitting being proportional to the applied magnetic field.[13] This splitting is thought to be a result of the spin up and spin down populations pairing separately.[14] The sound attenuation peak associated with the A transition splits into two separate peaks corresponding to the A_1 and A_2 transitions.[15] This splitting is clearly evident in the $\theta = 0$ attenuation data for 4.0 and 7.5 kOe shown in Fig. 3. The peaks obtained at $\theta = 0$ are highly reproducible and the attenuation data are rather smooth in this direction. The

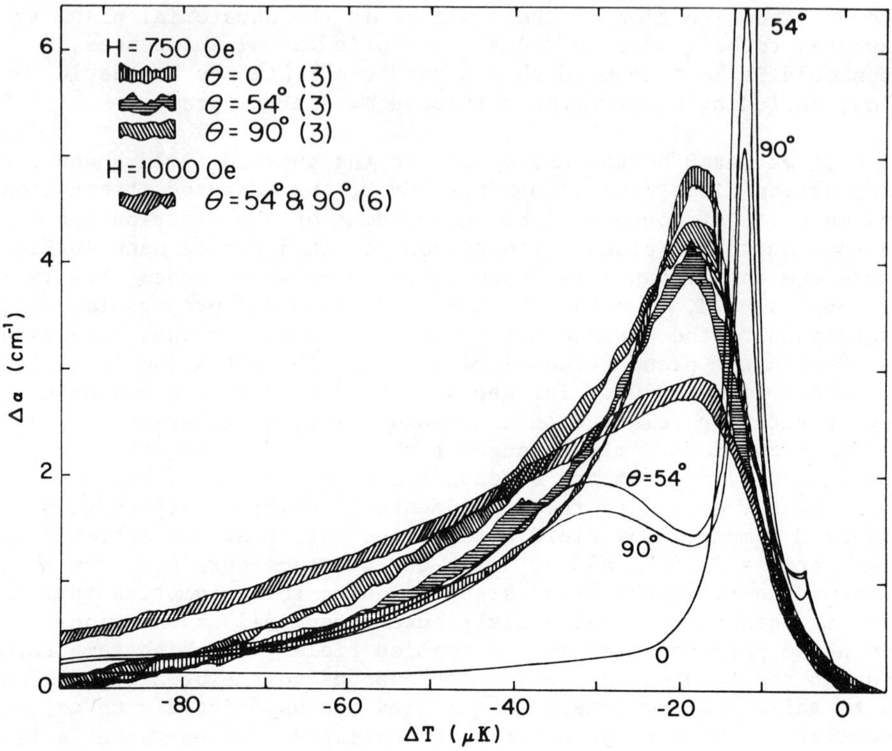

Figure 2. The A transition attenuation peaks at low field: $\Delta\alpha = \alpha - \alpha(T_A)$ vs $\Delta T = T - T_A$. Each of the shaded bands indicates the range of values for a particular orientation. The number of experimental curves represented by each band is noted in parentheses. The solid lines are based on theory (see Ref. 10); the peak of the truncated $\theta = 0$ curve lies at $\alpha = 11.5\ \text{cm}^{-1}$. The 1000-Oe (0.1 T) data are examples of a distinctly different peak shape, discussed in Ref. 8, which displays no anisotropy.

reproducibility of the data in the $\theta = 0$ direction is consistent with the interpretation given in Fig. 1b. The $\vec{\ell}$ vector is always perpendicular to $\hat{\kappa}$ no matter what orientation $\vec{\ell}$ assumes in the plane perpendicular to $\vec{H}$ (and $\hat{\kappa}$ for $\theta = 0$). It is reasonable that the attenuation of compressional sound in any direction orthogonal to $\vec{\ell}$ is the same because of the symmetry of the gap in the plane perpendicular to $\vec{\ell}$.

The situation is quite different for data taken at $\theta = 90°$, also given in Fig. 3 for H = 4.0 and 7.5 kOe. These attenuation

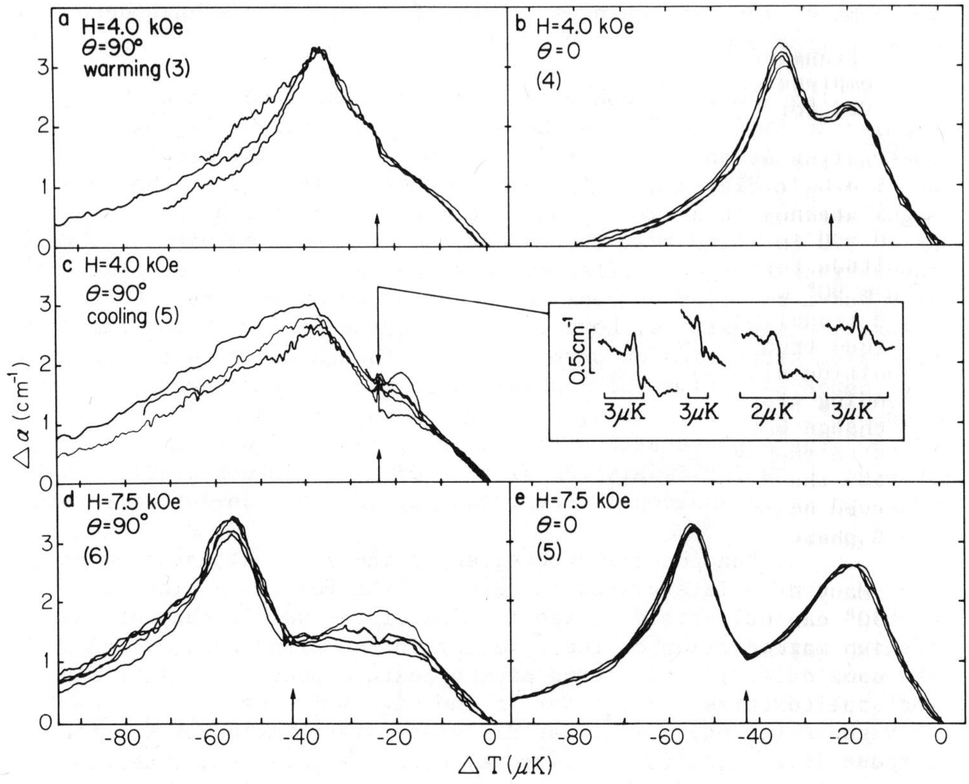

Figure 3. A transition attenuation peaks for various magnetic field strengths and orientations. $\Delta\alpha = \alpha - \alpha(T_A)$; $\Delta T = T - T_A$. The number of experimental curves included is shown in parentheses for each case; the ΔT corresponding to T_{A_2} is indicated by an arrow. Except for (a) and (c), each plot includes data taken both while warming and while cooling. The inset shows in more detail, from the H = 4 kOe raw data, some examples of the A_2 "oscillation." $T_{A_1} \simeq 2.8$ mK; $\alpha(T_{A_1}) \simeq 2$ cm^{-1}.

curves are characterized by fluctuations with periods typically on the order of 10 seconds, less well defined A_1 peaks, considerable variation between different curves taken at the same value of the field and strong reproducible warming-cooling asymmetry in the θ = 90°, 4.0 kOe attenuation peaks. Another new phenomenon is found at the high temperature edge of the A_2 attenuation peak for θ = 90°. Observed at all fields $\geq$ 2 kOe, this new feature is a narrow (< 3 μK), reproducible oscillation of attenuation with temperature, about 0.5 cm^{-1} in maximum amplitude (see inset in Fig. 3). Finally, large transient effects in the $\theta = 90°$ data are observed accompanying compression rate changes. None of these effects have been observed for $\theta = 0$.

As the sample is cooled from the A phase into the B phase, a small drop ($\sim 0.1\ cm^{-1}$) in the attenuation occurs. Fig. 4 shows these attenuation steps for $\theta = 0$ and $\theta = 90°$. The steps shown for $\theta = 0$ in Fig. 4 are highly reproducible and are of the same magnitude for both warming and cooling through the B transition. At $\theta = 90°$ we find that the attenuation increase on warming through the B transition may differ by more than a factor of two from the previous transition on cooling at the same temperature. Cooling transition signatures, however, agree quite well with the immediately preceding transition on warming. Even though the size of the steps may change each time the A phase is formed, the attenuation always drops to the same value upon returning to the B phase. Thus for $\theta = 90°$ the A phase attenuation is preparation dependent. The observed behavior is also consistent with the interpretation that the B phase is isotropic, corresponding to a Balian-Werthamer state.[16]

Many of the phenomena observed in the sound attenuation at $\theta = 90°$ can be interpreted in terms of the following picture: At high magnetic fields large regions of the sample chamber have the same orientation of the $\vec{\ell}$ vector in the plane perpendicular to the applied field. In terms of the texture model, we might say that the textures are coarser at the higher fields. Each time the A phase is formed, a somewhat different distribution of $\vec{\ell}$ orientations (i.e. a different texture) might be obtained, thus causing large differences in observations made on successive passages into the A phase.

Since the direction of $\vec{\ell}$ is influenced by flow (including supercurrents induced by thermal gradients), cell vibrations or stray heat inputs might change the orientation of $\vec{\ell}$ in different parts of the sample, thereby giving rise to fluctuations of the sound attenuation. Only at higher fields where coarser textures are present will this effect be significant. At low fields (< 1 kOe) the fluctuations become unobservable in our apparatus. The fluctuations in sound attenuation also may be manifestations of the orbit waves proposed by Anderson.[17]

In Fig. 5, we show raw data directly from chart recorder traces taken at H = 1.0 T (10 kOe) for $\theta = 0$ and $\theta = 90°$ during the attenuation measurements. The vertical axis is the detected and integrated voltage corresponding to the entire first received pulse. The difference in smoothness of the $\theta = 0$ data as compared with the 90° data is a dramatic illustration of the fluctuation effect. A new feature can be seen on the leading edge of the A_2 peak in the $\theta = 0$ data. This small feature is not understood at this time; it

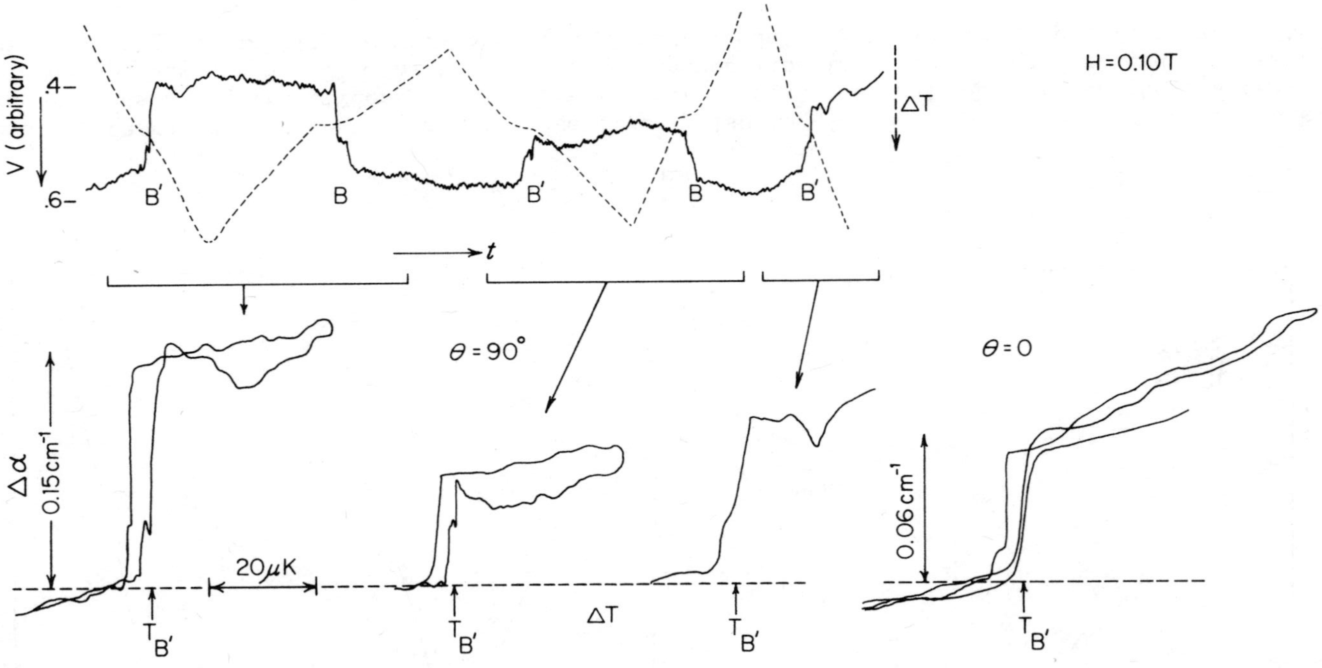

Figure 4. Top: Raw voltage vs time and pressure (temperature) vs time data (top) and the corresponding attenuation steps at the B transition for θ = 90°. The variations are large and show a preparation dependence. Bottom: Attenuation steps at the B transition for θ = 0°. The variations are small. Transitions are labeled B' when they occur on warming.

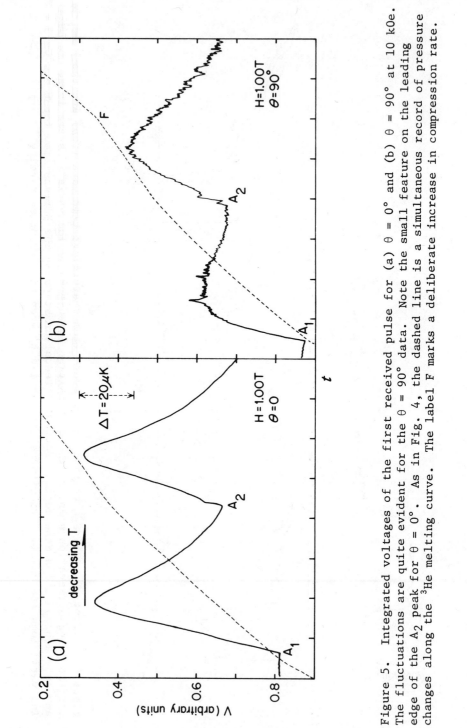

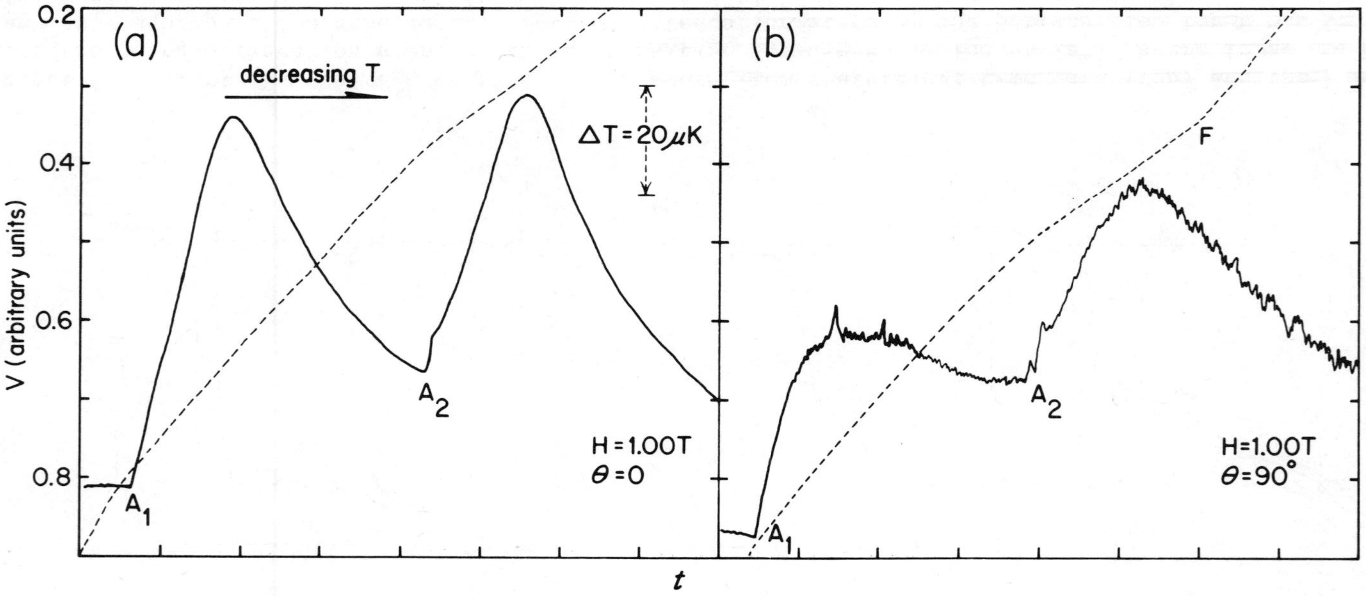

Figure 5. Integrated voltages of the first received pulse for (a) θ = 0° and (b) θ = 90° at 10 kOe. The fluctuations are quite evident for the θ = 90° data. Note the small feature on the leading edge of the A_2 peak for θ = 0°. As in Fig. 4, the dashed line is a simultaneous record of pressure changes along the ^{3}He melting curve. The label F marks a deliberate increase in compression rate.

may be related to the somewhat broader pair-breaking cusp of the theory or to the somewhat narrower A_2 oscillation seen at 90°.

Using a rather unusual phase-sensitive technique, we also have measured changes in the velocity of compressional sound to a few parts in 10^5 over a wide range of temperature and magnetic field strength in the superfluid phases of ^{3}He. As with attenuation all our velocity measurements were made at the ^{3}He melting pressure with 20 Mhz longitudinal sound propagating horizontally across a 4.14 mm sample.

An oscillator provided a 20 MHz cw signal which was gated by a phase-locked 0.5 μs square wave to produce an ultrasonic pulse for the transmitter. The same oscillator provided a reference signal for phase-sensitive detection of the received pulse by a Matec model 615 tuned receiver. Pulse repetition rates of 5 s^{-1} were typical. The oscillator was detuned slightly from the resonant frequency of the receiving quartz transducer until two or three zero crossings could be seen on each phase-detected pulse. Each of these pulses was stored in a Biomation model 8100 transient recorder for examination by a boxcar integrator during most of the period between transmissions. A variable delay on the boxcar integrator was regulated to keep its gate centered on a zero crossing early in the phase-detected pulse. Because phase changes $\Delta\phi$ in the beat pattern were equal to phase changes in the received rf pulses, changes in this gate delay Δt_g were proportional to changes in the zero sound transit time. Thus the voltage analog of the delay was recorded as a function of temperature, with shifts to adjacent zero crossings being made occasionally as the beat pattern moved across the detected pulses. Such $\Delta\phi = \pi$ shifts ($|\Delta t_g| \equiv t_\pi$), along with the known zero sound frequency, served to calibrate the velocity measurement. Linearity was checked by simultaneously monitoring two adjacent zero crossings as the sample underwent an A transition. A sample of the raw velocity data, including A and B transitions, is shown in Fig. 6.

For our frequency and path length and an absolute velocity c_0 of approximately 420 m s^{-1}, changes in the transit time are given by

$$\Delta t/t = 2.54 \times 10^{-3} \, \Delta t_g/t_\pi \ .$$

For $\Delta t/t \ll 1$, of course, $\Delta c_0/c_0 \simeq \Delta t/t$.

Because the compressional cooling technique restricted our observations to the melting curve, there is a variation of velocity with pressure which must be distinguished from the temperature dependence we wish to study. Since both c_1 and c_0 are expected to be independent of temperature in the normal liquid[18] we can obtain there accurate data for $\partial c_1/\partial P$ above the zero sound-first sound

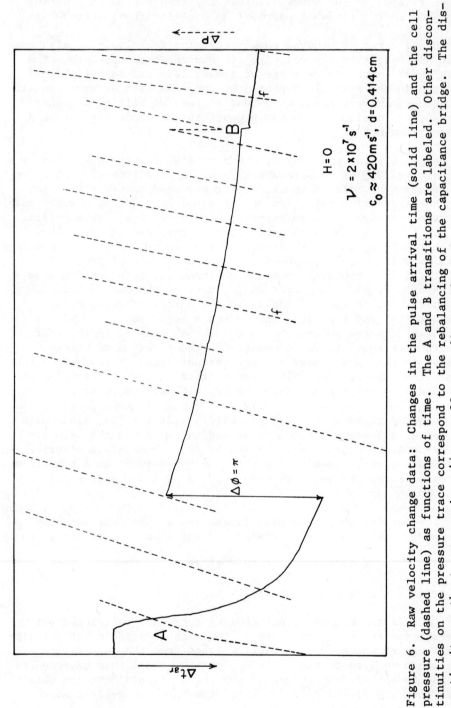

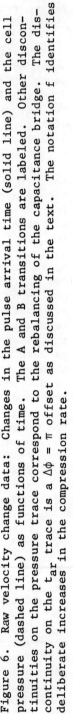

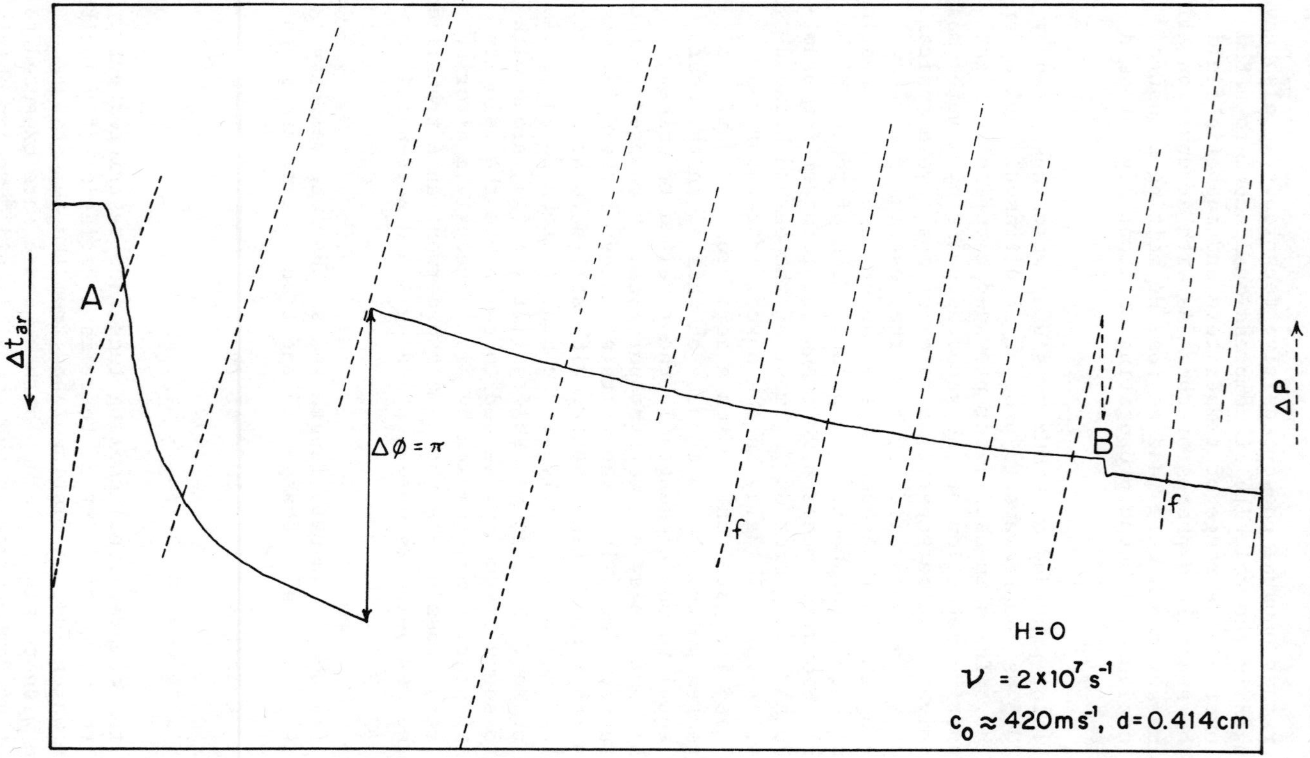

Figure 6. Raw velocity change data: Changes in the pulse arrival time (solid line) and the cell pressure (dashed line) as functions of time. The A and B transitions are labeled. Other discontinuities on the pressure trace correspond to the rebalancing of the capacitance bridge. The discontinuity on the t_{ar} trace is a $\Delta\phi = \pi$ offset as discussed in the text. The notation f identifies deliberate increases in the compression rate.

transition temperature and for $\partial c_o/\partial P$ below. For our frequency this transition was centered at 9.5° mK, where an attenuation maximum of 10 cm^{-1} was observed.

In the temperature interval 14-20 mK we obtain a value for $\frac{\partial c_1}{\partial P}$ of 5 m s^{-1} bar^{-1}. Using the values obtained by Halperin et al.[9] for $\frac{\partial P}{\partial T}$ as a function of T along the melting curve, we then can extrapolate Δc_1 to lower temperatures from a known value at some T_o:

$$\Delta c_1(T_i) = \Delta c_1(T_o) + \frac{\partial c_1}{\partial P} \int_{T_o}^{T_i} \frac{\partial P}{\partial T}\, dT \; .$$

In Fig. 7 we show typical data for velocity changes in magnetic fields of 0 and 0.20 T, below 20 mK. (The H = 0 melting curve temperature scale has been used in both cases.) The two dashed lines are the extrapolation described above and a similar extrapolation from $\Delta c_o(T_A) \equiv 0$ to higher temperatures, assuming that

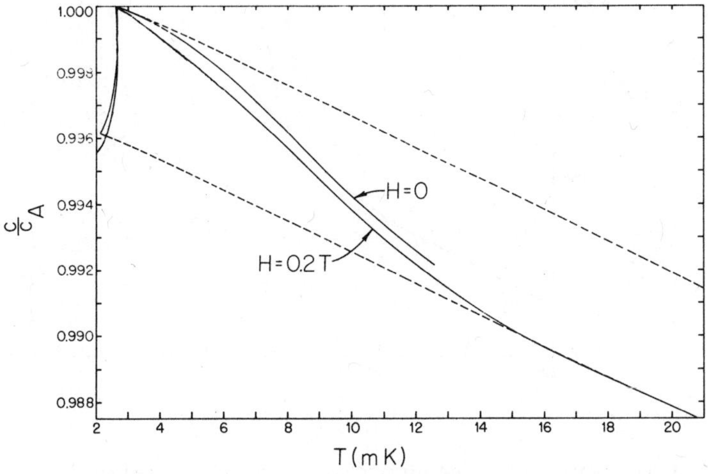

Figure 7. The first sound-zero sound transition. The ratio of compressional sound velocity to that at the A transition, as a function of temperature along the ^{3}He melting curve. The H = 0.2 T (2.0 kOe) data were taken at θ = 90°. The extrapolations shown as dashed lines are discussed in the text.

$\frac{\partial c_0}{\partial P} = \frac{\partial c_1}{\partial P}$. Notice that the first sound to zero sound velocity transition is effectively complete before the A transition intervenes.

As may be seen in the raw data of Fig. 6, both the A and B transitions are accompanied by well-defined signatures in the velocity. Just below T_A the velocity begins a drop of some 0.4-0.5%, comparable in magnitude to the first sound-zero sound transition.[19] At the B transition there is a further, abrupt decrease in velocity--roughly 0.01%. Both these effects also have been studied recently at lower pressures by Ketterson and co-workers.

Figure 8 includes details of the A transition signatures for three different values of magnetic field strength. We find that the H = 0 velocity data are much less reproducible than is the case for $H \geq 0.1T$. The H = 0.75T (7.5 kOe) curve shows evidence of the $A_1 - A_2$ splitting. At H = 1.00T the two transitions are distinct

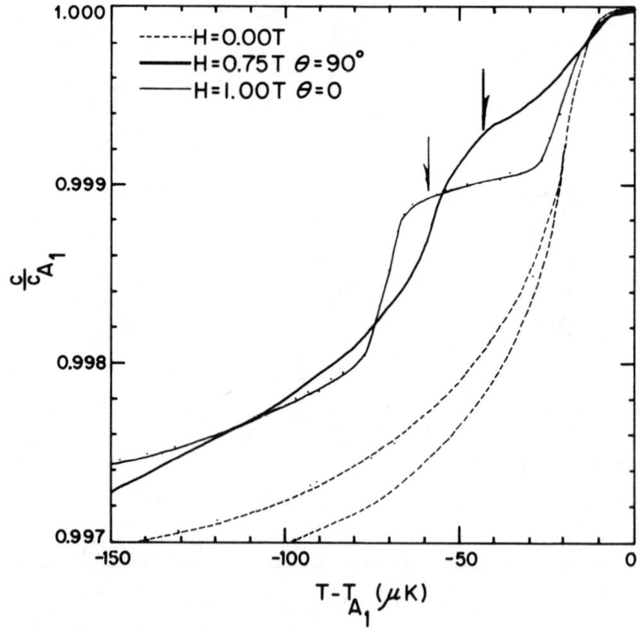

Figure 8. Temperature dependence of the compressional sound velocity at the A transition. For H = 0, the characteristic velocity change is strikingly non-reproducible. In a field of 0.75T (7.5 kOe) the $A_1 - A_2$ splitting causes a definite inflection of the velocity curve. For H = 1.0T the two transitions are nicely resolved. Arrows indicate the temperature of A_2, as obtained from attenuation measurements, in each case.

and well resolved in velocity.

Examples of transitions between the A and B phases are shown in Fig. 9. The initial A-to-B transition is supercooled in each case, with several subsequent transitions shown at the well-defined transition temperature $T_{B'}$. For $\theta = 0$ (propagation parallel to the applied magnetic field) all the transitions at $T_{B'}$ are between the same two values of velocity. In the $\theta = 90°$ case, however, the upper velocity (the A phase value) seems to be preparation dependent in much the same way as the attenuation. The present experiments have found no velocity change at B that approaches the crudely-measured 0.5% difference seen in some earlier experiments for 10 MHz sound propagating vertically.[15] A velocity change of such magnitude would correspond to a phase change of about 2π in the present experiment (see Fig. 6)

We wish to thank W. J. Gully, R. C. Richardson, J. D. Reppy, and R. A. Buhrman for their help in experimental matters and

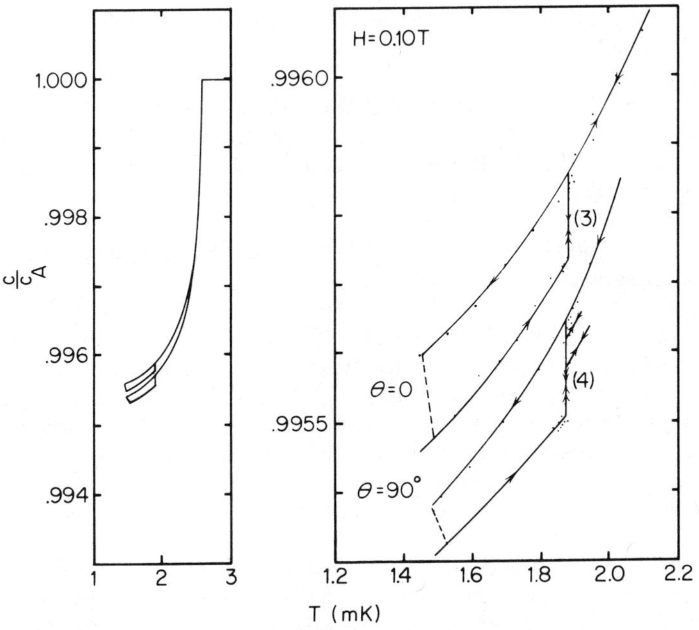

Figure 9. Temperature dependence of the compressional sound velocity at the B transition. Left: overall velocity change below the A transition. Right: detail showing orientation-dependent effects at the B transition. The numbers in parentheses indicate how many transitions are included at the same temperature; arrowheads indicate the time sequence of the events.

J. W. Serene, P. G. de Gennes, P. Wölfle, and J. W. Wilkins for helpful discussions of the theory. We are grateful to H. A. Fairbank and E. K. Riedel for a critical reading of the manuscript.

References

(1) D. D. Osheroff, W. J. Gully, R. C. Richardson and D. M. Lee, Phys. Rev. Lett. 29, 920 (1972).

(2) A. J. Leggett, Ann. Phys. (New York) 85, 11 (1974).

(3) P. W. Anderson and P. Morel, Phys. Rev. 123, 1911 (1961).

(4) P. W. Anderson and W. F. Brinkman, Phys. Rev. Lett. 30, 1108 (1973).

(5) P. G. de Gennes, Proceedings of the Nobel Symposium 24, Collective Properties of Physical Systems, edited by B. Lundqvist and S. Lundqvist (Academic Press, New York, 1973), p. 112.

(6) V. Ambegaokar, P. G. de Gennes and D. Rainer, Phys. Rev. A 9, 2676 (1974).

(7) P. G. de Gennes and D. Rainer, Phys. Lett. 46A, 429 (1974).

(8) D. T. Lawson, H. M. Bozler and D. M. Lee, Phys. Rev. Lett. 34, 121 (1975).

(9) W. P. Halperin, C. N. Archie, F. B. Rasmussen, and R. C. Richardson, to be published.

(10) P. Wölfle, Phys. Rev. Lett. 30, 1169 (1973) and to be published; H. Ebisawa and K. Maki, Prog. Theor. Phys. 51, 337 (1974); J. W. Serene, thesis, Cornell University, 1974 (unpublished), and to be published. We have used the expressions derived by Serene in our calculations.

(11) P. Wölfle, private communication.

(12) J. B. Ketterson (these proceedings); Pat R. Roach, B. M. Abraham, M. Kuchnir, and J. B. Ketterson, preprint; Pat R. Roach, B. M. Abraham, P. D. Roach, and J. B. Ketterson, preprint.

(13) W. J. Gully, D. D. Osheroff, D. T. Lawson, R. C. Richardson and D. M. Lee, Phys. Rev. A 8, 1633 (1973).

(14) V. Ambegaokar and N. D. Mermin, Phys. Rev. Lett. 30, 81 (1973).

(15) D. T. Lawson, W. J. Gully, S. Goldstein, R. C. Richardson and D. M. Lee, Phys. Rev. Lett. 30, 541 (1973), and J. Low Temp. Phys. 15, 169 (1974).

(16) R. Balian and N. R. Werthamer, Phys. Rev. 131, 1553 (1963).

(17) P. W. Anderson, Phys. Rev. Lett. 30, 368 (1973).

(18) W. R. Abel, A. C. Anderson and J. C. Wheatley, Phys. Rev. Lett. 17, 74 (1966).

(19) D. N. Paulson, R. T. Johnson, and J. C. Wheatley, Phys. Rev. Lett. 30, 829 (1973).

SOUND PROPAGATION IN NORMAL AND SUPERFLUID ^{3}He*

J. B. Ketterson

Northwestern University, Evanston, Illinois 60201 and

Argonne National Laboratory, Argonne, Illinois 60439

and

Pat R. Roach, B. M. Abraham, and Paul D. Roach

Argonne National Laboratory, Argonne, Illinois 60439

Abstract: We present measurements of the pressure dependence of the attenuation and velocity of sound in the hydrodynamic regime and in both the normal and superfluid zero sound regime of ^{3}He. The velocity and attenuation were studied at a frequency of 20.24 MHz and at pressures of 17.00, 21.00, 21.50, 21.80, 22.00, 23.17, 26.00, and 28.00 bar in the zero sound regime. In the hydrodynamic region the temperature dependence of the attenuation was studied at 5.48 and 10.02 MHz and at pressures of 0.69, 1.38, and 2.76 bar. At a pressure of 29.3 bar and frequency of 20.24 MHz the transition from hydrodynamic to zero sound behavior was studied for both the velocity and attenuation. In addition, the anisotropy of the velocity and the attenuation as a function of the angle between the direction of an applied external magnetic field and the sound propagation direction was observed in the superfluid A phase at a pressure of 26.0 bar; no anisotropy was observed in the B phase at 21.0 bar. The observed behavior associated with a collective excitation of the order parameter in the B phase is shown to be qualitatively in agreement with theoretical predictions. At pressures slightly above the polycritical point some unexplained structure is observed in the velocity near the AB transition.

*Work performed under the auspices of the U. S. Energy Research and Development Administration.

I. INTRODUCTION

The propagation of ultrasound is a powerful probe for studying the liquid phases of ^3He. The nature of the sound propagation depends on the magnitude of $\omega\tau$ where ω is the sound frequency and τ is a viscous relaxation time of the liquid. When $\omega\tau << 1$ (which is equivalent to the sound period being much longer than τ) sound propagates as the usual density wave familiar in hydrodynamics (first sound); when $\omega\tau >> 1$ density waves propagate as a collective mode known as collisionless or zero sound.[1-4]

At high temperatures the attenuation is given by the hydrodynamic expression[5]

$$\alpha = \frac{\omega^2}{2\rho c_1^3}\left[\frac{4}{3}\eta + \zeta + \frac{\kappa}{C_p}\left(\frac{C_p}{C_v} - 1\right)\right] \quad (1a)$$

where ρ, c_1, η, ζ, κ, C_p, and C_v are the density, first sound velocity, first viscosity, second viscosity, thermal conductivity, and heat capacities at constant pressure and constant volume, respectively. For ^3He ζ is negligible and below 500 mK $C_p \cong C_v$; the attenuation is then given by

$$\alpha_1 = \frac{2\omega^2\eta}{3\,c_1^3} \quad (1b)$$

and the attenuation becomes a direct measure of the viscosity. At much lower temperatures ^3He behaves as a Fermi liquid. In this limit the first sound velocity can be derived from the relation[1]

$$F_o = \frac{3mm^*c_1^2}{p_F} - 1 \quad (2)$$

where m is the mass of a ^3He atom, p_F is the Fermi momentum ($p_F = \hbar(3\pi^2\frac{\rho}{m})^{1/3}$) and F_o is a Landau parameter. The effective mass, m^*, is related to F_1, another Landau parameter, through Eq. (3) and can be evaluated from the specific heat.

$$\frac{m^*}{m} = 1 + \frac{F_1}{3} = \frac{C_V(\text{REAL})}{C_V(\text{IDEAL})} \quad (3)$$

As the Fermi liquid regime is approached the viscous relaxation time and the viscosity take up an asymptotic T^{-2} dependence; thus the attenuation behaves as $\alpha_1 \propto \omega^2/T^2$. At sufficiently low temperatures the condition $\omega\tau > 1$ will be met and a transition to the zero sound mode will occur. The expressions for the zero sound velocity, c_o, and attenuation, α_o, are lengthy[3,6,7] and will not be given here.

Qualitatively, upon passing into the zero sound mode the sound velocity changes quickly to a greater value in the vicinity of $\omega\tau = 1$, and the attenuation passes through a maximum. The attenuation of zero sound, α_0, follows a T^2 law and is independent of frequency. At still lower temperatures direct excitation of quasiparticles can occur and the attenuation has the form $\alpha = \alpha_0(1 + \frac{\hbar}{2\pi k_B T})$); this "quantum limit" effect has not been observed to date.

At temperatures below about 3×10^{-3}K, ^{3}He makes a transition into a superfluid phase;[8-10] at still lower temperatures, depending on the pressure, another superfluid phase exists. The pressure, temperature, and magnetic field dependence of the transition temperatures has been studied in some detail. Figure 1 shows the phase diagram in the P-T plane as given by the group at Helsinki University of Technology.[13] A line of second order phase transitions separates the normal liquid from the superfluid. A line of first order phase transitions, which intersects the line of second order transitions at a point known as the polycritical point (PCP), separates a superfluid region in the P-T plane at higher temperatures and pressures known as the A phase; the remainder of the superfluid region of the phase diagram is known as the B phase.

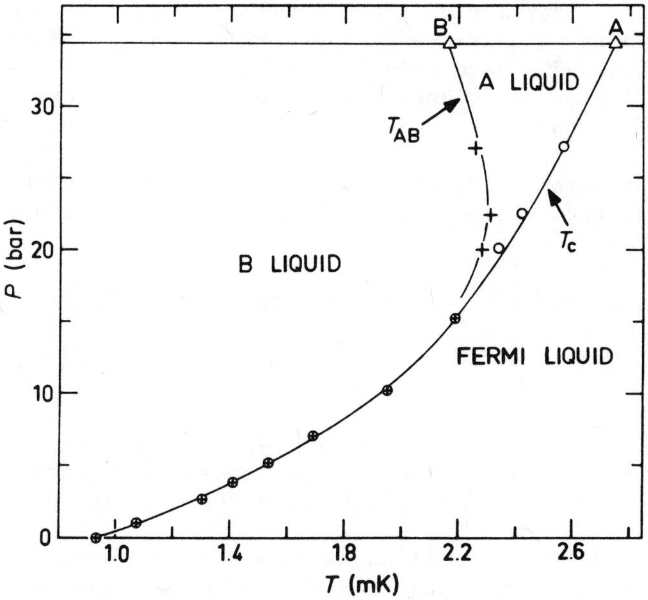

Figure 1. The phase diagram of superfluid ^{3}He in the P-T plane as given by Ahonen, et al.[13]

Both of the phases are believed to be Bardeen-Cooper-Schreiffer states in which quasiparticles of opposite momenta are paired; the unique feature of the ^{3}He superfluids in that the pairing occurs in a triplet (S = 1) state which, by the Pauli principle, must have an odd angular momentum. Anderson and Morel[14] have proposed a state in which only the spin up and spin down components of the triplet are paired; a modification of this type of state proposed by Anderson and Brinkman,[15] known as the ABM, or axial, or anisotropic state, is believed to correspond to ^{3}He A; the most convincing evidence for this identification comes from the interpretation of the NMR experiments.[8,16] For this phase the energy gap is anisotropic being zero along the angular momentum axis, ℓ, and largest on the equator. Balian and Werthamer[16a] proposed a phase in which all three components of the triplet state take part in the pairing; this state, known as the BW or isotropic state, is believed to correspond to the B phase of superfluid ^{3}He and here the energy gap is isotropic.

Theoretical aspects of sound propagation in superfluid ^{3}He have been studied by Wölfle,[17] Ebisawa and Maki,[18] Maki,[19] and Serene.[20] All of the published work to date has neglected the effects of quasiparticle lifetime. Using methods developed by Wölfle[17] which incorporate pairing theory into Landau transport theory, Serene[20] has calculated the attenuation and velocity in both the A and B phases. Using thermal Green's function techniques the A phase and the B phase have been studied by Ebisawa and Maki[18] and by Maki,[19] respectively. The results of both methods are essentially identical. Wölfle[20a] and Maki and Ebisawa[20b] have recently extended the theory to include lifetime effects. They have made some detailed numerical calculations for the B phase.

The anisotropic nature of the A liquid leads to an anisotropy in the sound propagation. Serene has written the anisotropy in the following form

$$\alpha = \alpha_{||}(T)\cos^4\theta + 2\alpha_c(T)\cos^2\theta\sin^2\theta + \alpha_{\perp}(T)\sin^4\theta \tag{4a}$$

$$\Delta c = \Delta c_{||}(T)\cos^4\theta + 2\Delta c_c(T)\cos^2\theta\sin^2\theta + c_{\perp}(T)\sin^4\theta \tag{4b}$$

where θ is the angle between the propagation direction, $\hat{q}$, and the axial-state-energy-gap axis, $\hat{\ell}$. The functions $\alpha_{||}(T)$, $\alpha_c(T)$, $\alpha_{\perp}(T)$, $\Delta c_{||}(T)$, $\Delta c_c(T)$, and $\Delta c_{\perp}(T)$ have been computed by Serene (his Eq.

(6.17); Fig. 13 (which will be discussed later) shows a plot of these quantities as a function of temperature for a pressure of 26.0 atm.[21] Two physical processes produce structure in the attenuation and velocity. The first of these is pair breaking which occurs when $\hbar\omega > 2\Delta$. The small sharp structure in the attenuation just below the transition temperature may result from pair breaking near the equator of the Fermi surface where there is a singularity in the density of states. The second physical process is the excitation of a collective mode of the energy gap (or order parameter) by the sound wave; there are two such modes for the A phase and one mode in the B phase. The large sharp peak in $\alpha_\perp(T)$ and the broad peak in $\alpha_c(T)$ arise from the effects of these two modes.

The dipole-dipole interaction energy favors the angular momentum vector $\hat{\ell}$ aligning in the plane perpendicular to the direction of an external magnetic field $\hat{H}$ and may lie anywhere in this plane (in the absence of boundary effects which may be negligible in the bulk liquid). This situation is shown pictorially in Fig. 2. Let us assume that a bulk sample of A liquid in a magnetic field is made up of many domains or textures of liquid whose $\hat{\ell}$ vectors lie randomly in the plane perpendicular to $\hat{H}$. Let ψ be the angle between $\hat{H}$ and $\hat{q}$ and let ϕ be the angle of $\hat{\ell}$ relative to the intersection of the plane formed by $\hat{H}$ and $\hat{q}$, and the plane perpendicular to $\hat{H}$. We then have $\cos\theta = \hat{\ell}\cdot\hat{q} = \cos\phi\sin\psi$; inserting this expression in Eqs. (4a) and (4b) and averaging over ϕ we obtain

$$\alpha = (1/8)(3\alpha_{||} + 2\alpha_c + 3\alpha_\perp)\sin^4\psi + (\alpha_\perp + \alpha_c)\cos^2\psi\sin^2\psi + \alpha_\perp\cos^4\psi \tag{5a}$$

$$\Delta c = (1/8)(3\Delta c_{||} + 2\Delta c_c + 3\Delta c_\perp)\sin^4\psi + (\Delta c_\perp + \Delta c_c)\cos^2\psi\sin^2\psi + \Delta c_\perp\cos^4\psi \tag{5b}$$

Note that Eqs. (4) and (5) have the same functional form for the angular dependence. By fitting the attenuation (or velocity) data to Eq. (5a) (or (5b)), which is a trigonometric polynomial with three coefficients, we can algebraically extract the three coefficients of Eq. (4a) (or (4b)). Thus, for sound propagation work, it may not be necessary to have a fully oriented sample. We note that the anisotropy predicted by Eq. (4) is identical to that

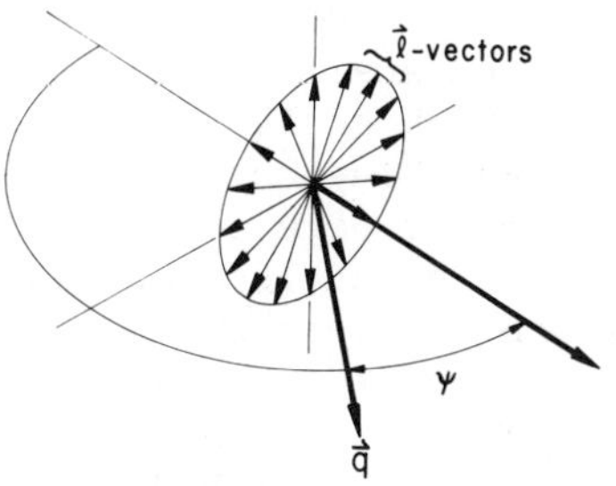

Figure 2. The "fan" of ℓ vectors for a given field direction $\hat{H}$.

observed in a smectic-A liquid crystal;[22] Eq. (5b) has been used (in the high frequency limit) to represent a cholesteric liquid crystal as a "twisted" nematic.[23]

The attenuation and velocity in the B phase were calculated by Serene[20] (his Eqs. 7.90 and 7.91) and by Maki[19] (his Eq. 34)). In the limit $h\omega << 4k_BT$ Serene and Maki found

$$\alpha = \frac{3\pi\omega\Delta^2}{5F_o ck_BT}\left[\frac{(\hbar^2\omega^2-4\Delta^2)^{1/2}}{5\hbar^2\omega^2-12\Delta^2}\theta(\hbar\omega-2\Delta)+\frac{\pi}{5\sqrt{6}}\delta(\hbar\omega-2\sqrt{3/5}\,\Delta)\right] \quad (6a)$$

$$\frac{\Delta c}{c} = \frac{3\pi\Delta^2}{5F_o k_BT}\,\frac{(4\Delta^2-\hbar^2\omega^2)^{1/2}}{12\Delta^2-5\hbar^2\omega^2}\,\theta(2\Delta-\hbar\omega) \quad (6b)$$

where Δ is the (isotropic) energy gap, θ and δ are a unit step function and Dirac δ-function, respectively, and F_o is the zeroth order Landau parameter; the remaining symbols have their usual meanings. Qualitatively these formulas have the following features: (1) a broad attenuation band between T_C and the temperature at which $h\omega = 2\Delta(T)$ (this is associated with the pair breaking for $h\omega > 2\Delta$), (2) a δ function peak in the attenuation at $\hbar\omega = 2\sqrt{3/5}\,\Delta(T)$ (arising from coupling to a collective mode of the energy gap), (3) no shift in the sound velocity for $h\omega > 2\Delta(T)$, and (4) a divergence with sign reversal in the sound velocity at $h\omega = 2\sqrt{3/5}\,\Delta(T)$.

II. EXPERIMENTAL TECHNIQUES

The measurements of the pressure dependence of the attenuation in the first sound regime were performed with a dilution refrigerator and a metallic sonic cell which has been described elsewhere.[24]

The very low temperature measurements were performed in an epoxy adiabatic-demagnetization cell; a schematic drawing is shown in Fig. 3. The cell was precooled through a specially designed mechanical heat switch which was an integral part of the epoxy mixing chamber of our ^{3}He-^{4}He dilution refrigerator. The cell and heat switch are described in detail elsewhere.[25] The heat switch was actuated by a bellows which could be expanded (switch closed) using pressurized liquid ^{4}He. The mechanical contact surfaces were 1 cm^2 in area and were formed from a 1 cm high stack of 200 1 cm^2 × 0.0025 cm Au foils which had been electron beam welded together along one edge, after which the welded edge was machined flat to form the contact surface. One of these foil brushes was sealed (around the edge of the welded surface) into the demagnetization cell and another into the mixing chamber. The resulting 400 cm^2 surface area of foils of each switch section provided contact to

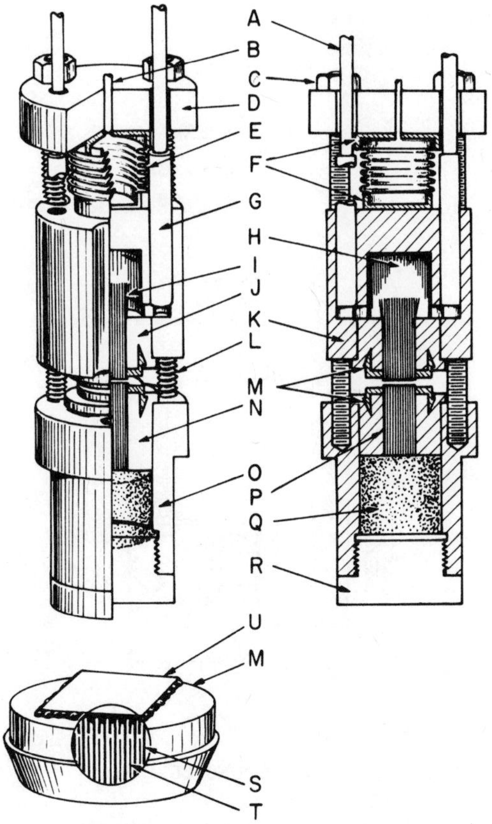

Figure 3. Adiabatic demagnetization cell: A) Mixing chamber input line; B) Bellows actuating capillary; C) Nylon nut; D) Epoxy support plate; E) ^{4}He bellows; F) Bellows end plates; G) Epoxy jacketed mixing chamber return line; H) Mixing chamber reservoir; I) Mixing chamber gold brush; J) Upper epoxy switch casting; K) Epoxy mixing chamber; L) Nylon support rods; M) Be-Cu epoxy to gold transition caps; N) Lower epoxy switch casting; O) Main epoxy cell body; P) Cell gold brush; Q) Paramagnetic powder and ^{3}He; R) Removable epoxy cell bottom; S) Individual gold foil; T) Lens paper; U) Welded gold switch surface.

the liquid in the demagnetization cell and in the mixing chamber. The demagnetization cell contained 23.2 grams of cerium magnesium nitrate (CMN) together with a measured volume of 2.4 cm^3 of ^{3}He (including the sonic chamber).

Figure 4 shows the sonic components which were contained in an

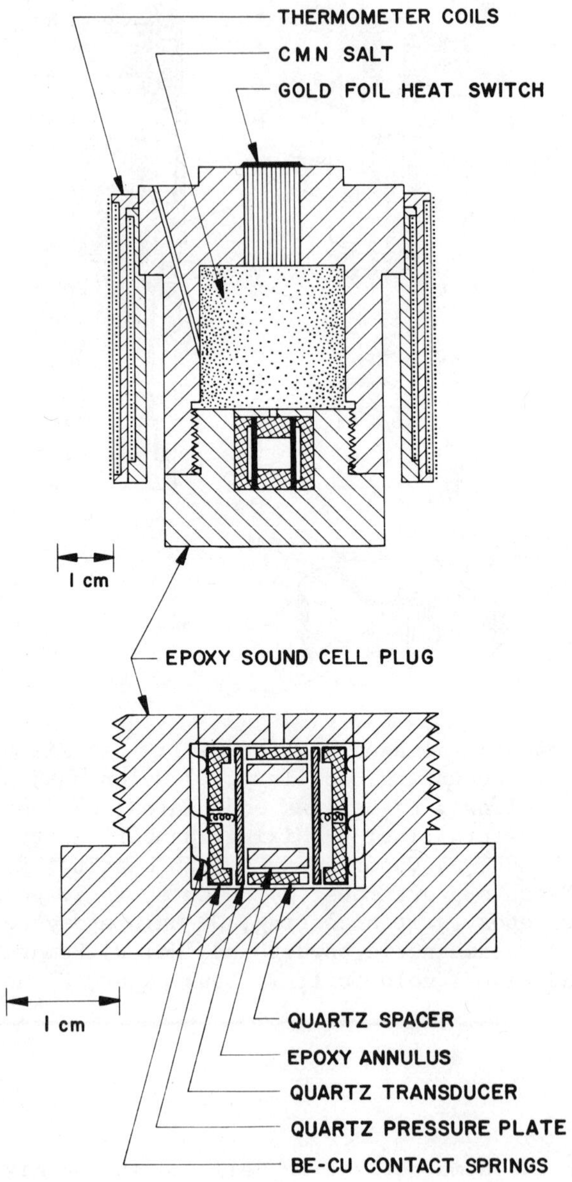

Figure 4. A schematic drawing of the demagnetization cell and epoxy sound plug.

epoxy plug which was screwed into the bottom of the demagnetization cell; the threads were sealed with Nonaq stopcock grease. The sound path was defined by a fused quartz annulus 0.6 cm long with an outside diameter of 1.27 cm and an inside diameter of 0.95 cm. An eopxy annulus slightly under 0.6 cm long was inserted into the quartz in order to reduce the volume of ^{3}He in the cell. The ends of the quartz spacer were crenelated (to minimize spurious transmission through the spacer and to allow thermal contact to the liquid) and ground optically flat (to minimize phase cancellation of the wave front). Appropriately machined epoxy pieces were fitted over the spacer to reduce further the quantity of ^{3}He; a hole together with various grooves allowed liquid thermal contact with the main reservoir. The ends of the spacer were capped by unloaded 20 MHz coaxially-plated x-cut quartz transducers; electrical contact and mechanical pressure between the transducers and the spacer were provided by gold-plated beryllium-copper spring contacts; the pairs of electrical leads to the input and output transducers were twisted to minimize cross talk.

The sound propagation axis, $\hat{q}$, was aligned perpendicular to the cryostat axis. A Helmholtz pair mounted on a rotatable base was located external to the cryostat such that the angle between $\hat{q}$ and $\hat{H}$ could be varied from 0-360°. Measurements of the temperature and field dependence (direction and magnitude) of the attenuation and velocity of 20.24 MHz sound were made using the phase comparison technique of Abraham, et al.[24] The repetition rate was typically 3-10 Hz and the peak to peak output level of the 5 μsec transmitter pulse was ∿ 100 mV. A stainless steel toepler pump system with an oil intensifier provided the hydrostatic pressures which were measured with a Texas Instruments precision quartz-Bourdon-tube gauge.

Temperatures were determined by monitoring the magnetic susceptibility of the CMN in the main body of the demagnetization cell using a standard mutual inductance bridge. The thermometer was calibrated at high temperature against the ^{3}He vapor pressure scale. An additive correction to this T^* scale was arrived at by comparison with other measurements of the phase diagram;[9,11] a T^2 variation of the zero sound attenuation was observed in the normal liquid at all pressures studied. It is important to take into account the effect of the field of the Helmholtz pair on the susceptibility of the cerium magnesium nitrate (CMN) used for thermometry. This was done by calibrating the CMN against the temperature dependence of the attenuation of zero sound in both the normal liquid and in the B phase. The attenuation peaks in the B phase were identical at different fields except for a temperature shift due to the change in the thermometer calibration. At pressures of 21.0 and 8.0 bar the attenuation was measured at H = 0, 9, 18, and 27 gauss; the susceptibility, χ, was found to follow Curie's law in the form $\chi = C(H)/(T + \Delta)$ for all fields where C(H) is a field dependent Curie constant and Δ is the additive correction to the temperature scale. The most consistent calibrations were obtained with the liquid in the B phase, presumably due to the better thermal equilibrium in the superfluid state.

In most of the experiments the cell was precooled to a temperature of 17-20 mK in a field of 1.7 kG. Providing that the temperature at the end of the previous demagnetization was below 3 or 4 mK, the time to cool back down to 20 mK following remagnetization was on the order of 30 hrs; it is felt that a better preparation of the contact flats of the heat switch could materially reduce this time. The demagnetization schedule was determined from an optimized computer simulation of the adiabatic demagnetization process[26] as well as experience with prior demagnetizations. The total demagnetization time was approximately 4 hours.

III. EXPERIMENTAL RESULTS

The data reported here will be divided into two categories: (A) those for the normal fluid in both the hydrodynamic and zero sound regimes, and (B) those primarily concerned with the superfluid regime.

A. Sound Propagation in the Normal Fermi Liquid Regime

As discussed in the introduction the attenuation in the $\omega\tau \ll 1$ limit is, to a high degree of accuracy, directly related to the viscosity through Eq. (1b). The viscosity of ^3He has been determined from measurements of capillary flow,[27,28] damping of a torsionally oscillating cylinder,[29-31,35] vibrating wire,[32] and the attenuation of ultrasound.[33,34]

Figure 5 shows our measurements of the attenuation of sound in liquid ^3He at the vapor pressure. The squares show our most recent measurements at 10.02 MHz; shown also are our earlier measurements at 5.48 MHz.[34] At very low temperatures, in the Fermi liquid regime, the attenuation, or equivalently the viscosity, should take up a T^{-2} dependence; the solid straight line shows such a temperature dependence where the proportionality coefficient has been taken from the work of Abel, Anderson, and Wheatley.[33] The remaining two lines are derived from the work of Betts, Osborne, Welber, and Wilks,[30] and from the recent measurements of Black, Hall, and Thompson.[32] Since in this regime the attenuation is proportional to the square of the frequency, ν, we can present both sets of data on the same curve by plotting α/ν^2; the vapor pressure data are shown in this form in Fig. 6. The line represents a smooth curve through the data points. It will be observed that the data from

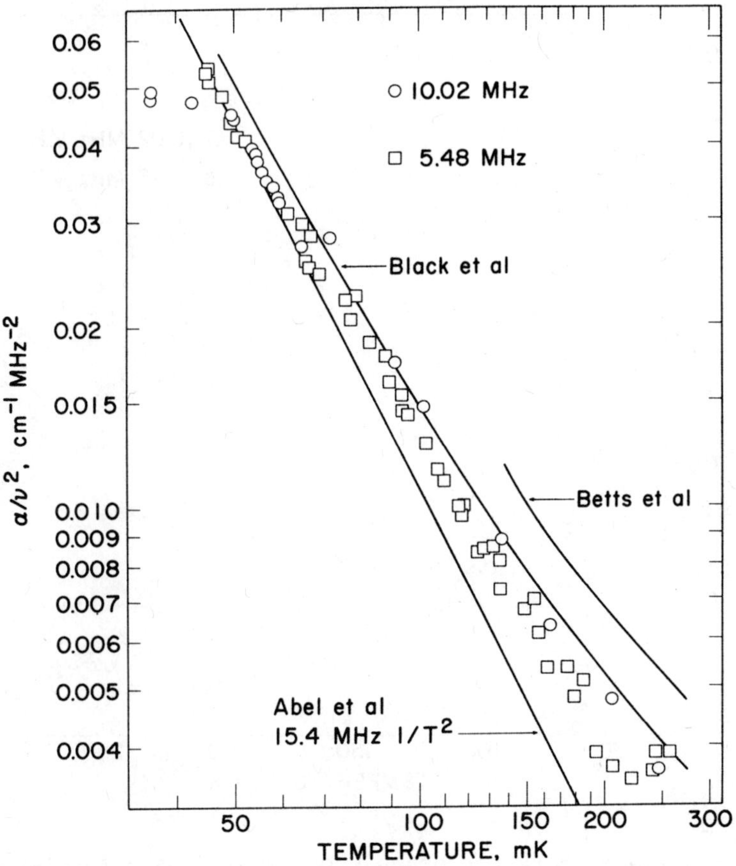

Figure 5. The temperature dependence of the attenuation of sound in liquid ^{3}He at the vapor pressure at frequencies of 10.02 MHz, and 5.48 MHz; shown also are the results of Black, et al.,[32] and Betts, et al.[30] For reference, a line showing the attenuation which would result from a T^{-2} dependence of η is shown where the coefficient was extracted from the data of Abel, et al.[33]

the two frequencies are essentially identical with the exception of the three lowest temperature points at 10.02 MHz; here the discrepancy could be due to poor thermal contact between the metallic cell and the external CMN thermometer.

As is well known, it is difficult to determine absolute attenuations using the pulse echo technique because of spurious attenuation mechanisms such as nonparallel transducers, beam

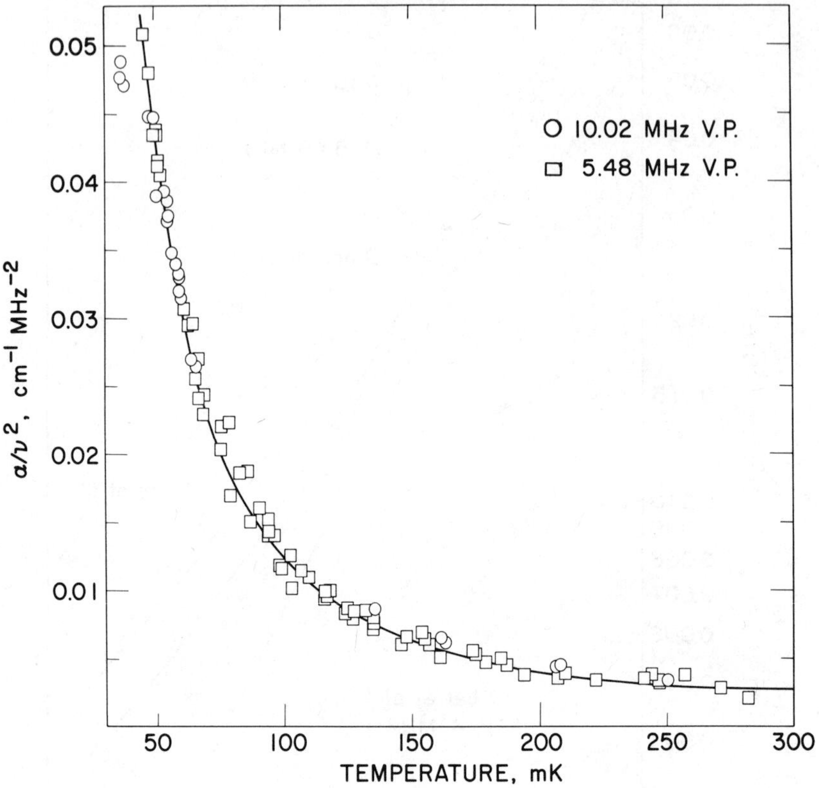

Figure 6. The temperature dependence of α/ν^2 as a function of temperature. The squares, □ , and circles, ○ , correspond to 5.48 and 10.02 MHz, respectively. The solid line is a smooth curve through the data.

spreading, extraction of signal for detection, etc.; the pulse comparison or the pulse interference methods do yield accurate measurements of the relative attenuation. Our data were normalized by computing the attenuation near 500 mK (where it is quite small) from the viscosity measurements of Betts, et al.[30] For 5.48 and 10.02 MHz this amounts to $\alpha \cong 0.4$ and 1.7 dB/cm, respectively.

Figure 7 shows our measurements of the temperature dependence of α/ν^2 under pressure; the pressures studied were 0.69, 1.38, and 2.76 bar, and the frequency was 10.02 MHz. The data under pressure were again normalized using the viscosity measurements under pressure; here we used the data of Betts, Keen, and Wilks.[35] Near

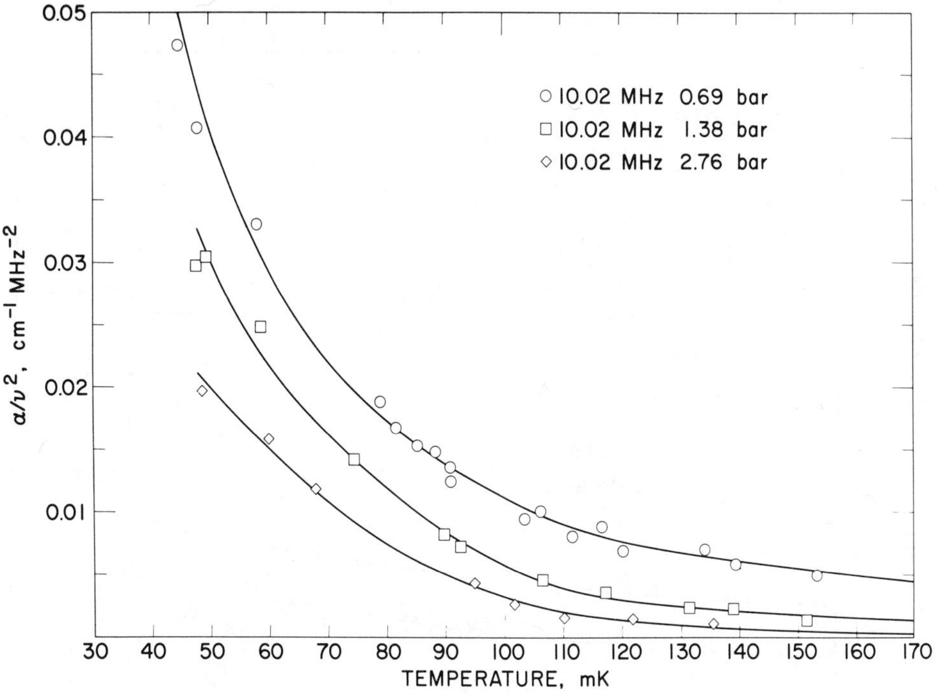

Figure 7. The temperature dependence of α/ν^2 at pressures of 0.69, 1.38, and 2.76 bar, respectively. The frequency was 10.02 MHz.

500 mK this amounts to $\alpha \cong 1.3$, 1.1, and 0.8 dB/cm for pressures of 0.69, 1.38, and 2.76 bar.

In no case do our attenuation data strictly follow a $1/T^2$ law. The deviations of the viscosity from the asymptotic Landau-Fermi liquid regime have been studied by Emery;[36] he has shown that the lowest order temperature dependent correction to the viscosity has the form

$$\left(\frac{1}{\eta T^2}\right)_0 - \left(\frac{1}{\eta T^2}\right)_T \propto T \quad , \tag{7}$$

where $(1/\eta T^2)_0$ denotes the zero temperature limit.

In Fig. 8 we show a plot of ηT^2 as a function of temperature for the vapor pressure and 0.2 and 0.69 bar; according to the above relation this product should be proportional to T at low temperatures. We observe that, although the data are consistent with this behavior,

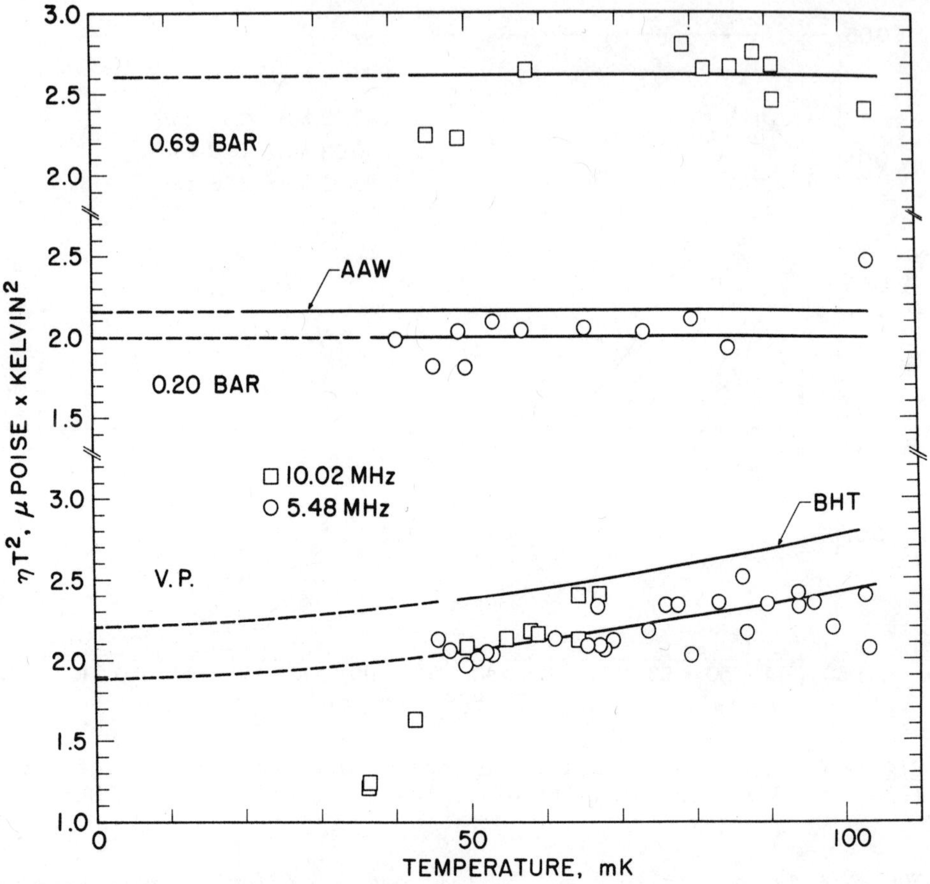

Figure 8. A plot of ηT^2 vs temperature for the vapor pressure, 0.2 bar and 0.69 bar; the frequency was 10.02 MHz in all cases.

the scatter is too large to extract the proportionality coefficient accurately; this plot does allow us to extract a value for $(\eta T^2)_0$.

Figure 9 shows the temperature dependence of the attenuation and velocity at a pressure of 29.3 bar and a frequency of 20.24 MHz. All regimes discussed in the introduction, with the exception of the quantum limit, are clearly visible in this figure. The attenuation data are plotted on a log-log scale so that a power law behavior of the attenuation with temperature corresponds to a straight line. The lines through the high and low temperature points show a T^{-2} and T^2 behavior; these two regions correspond to the hydrodynamic and zero sound regimes, respectively, and a transition between the

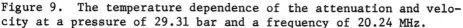

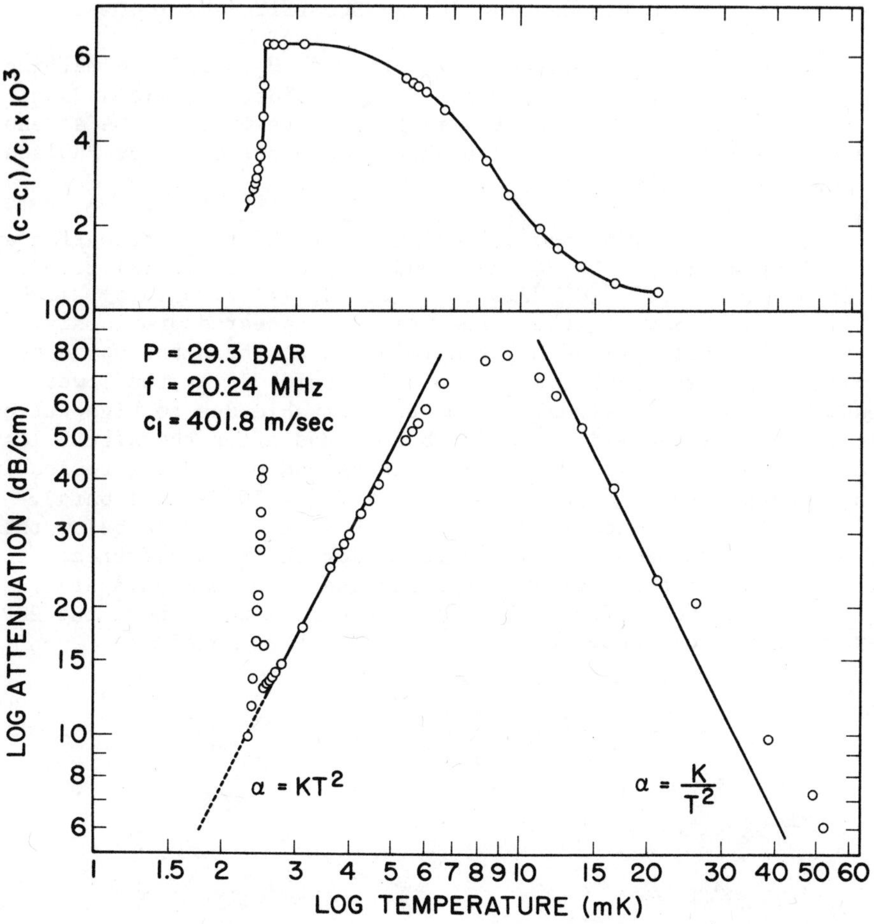

Figure 9. The temperature dependence of the attenuation and velocity at a pressure of 29.31 bar and a frequency of 20.24 MHz.

two occurs in the vicinity of 9.0 mK. The sharp peak near the low temperature extreme of the data is associated with the superfluid transition. The upper part of Fig. 9 shows the temperature dependence of the velocity. Note that it is essentially temperature independent except in the vicinity of 9 mK where the velocity increases on going from the hydrodynamic to the zero sound regime; in addition the velocity falls rapidly at the superfluid transition and, at very low temperature, approaches the hydrodynamic velocity.

B. Sound Propagation in the Superfluid Regime

The attenuation of sound in superfluid ^3He along the melting curve was first studied by Lawson, Gully, Goldstein, Richardson, and Lee.[37] Measurements off the melting curve of both the attenuation and velocity in the A and B phases were performed by Paulson, Johnson, and Wheatley.[38]

Figures 10a - 10h show our measurements of the attenuation and velocity of sound for 20.24 MHz sound at pressures of (a) 28.00, (b) 26.00, (c) 23.17, (d) 22.00, (e) 21.80, (f) 21.50, (g) 21.00, (h) 17.00 Bar. Some of these results have appeared previously.[38a] The velocity data were normalized relative to the first sound velocity (measured at 60 mK) in the normal liquid.[34] At the lower pressures, where the signal became unobservable due to high attenuation, the difference in velocity before and after the signal loss is uncertain to within multiples of the sound period (corresponding to an uncertainty in $\Delta c/c$ of multiples of 3×10^{-3} at 21 bars). This uncertainty at the attenuation peak in the B phase below the PCP was settled unambiguously by requiring that the difference in velocity between the normal liquid and the B-phase liquid at $\sim$ 2.2 mK be a smooth function of pressure from well above the PCP where we have continuous data to below the PCP. A similar

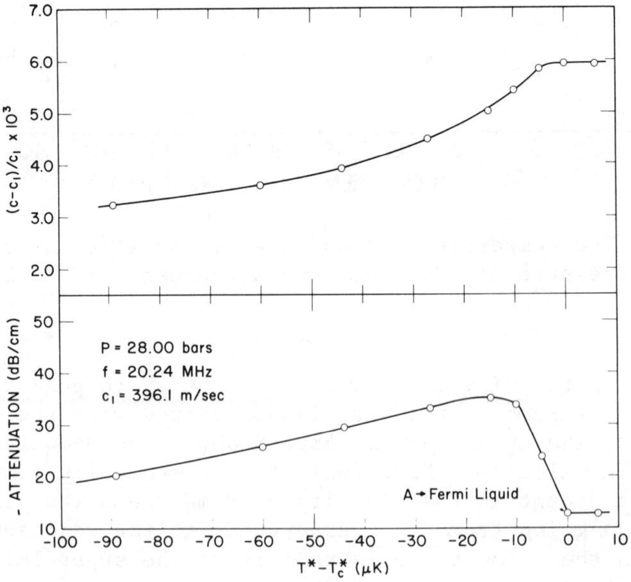

Figure 10. The attenuation and velocity of sound as a function of temperature at a frequency of 20.24 MHz: (a) 28.00 bar.

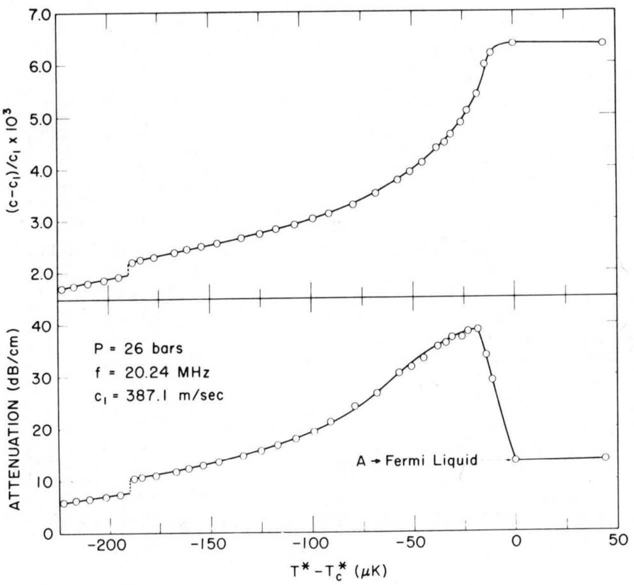

Figure 10. The attenuation and velocity of sound as a function of temperature at a frequency of 20.24 MHz: (b) 26.00 bar.

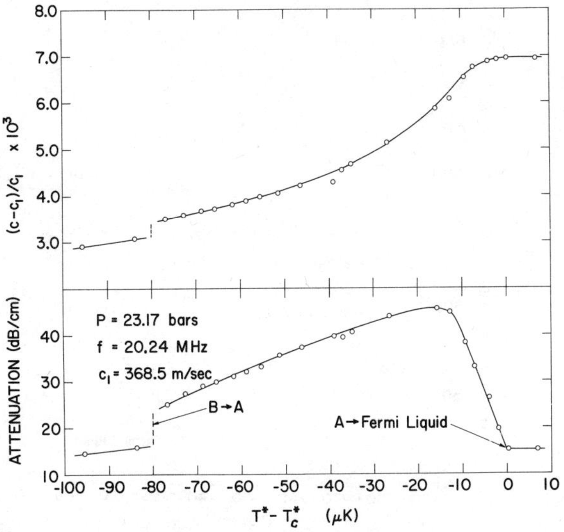

Figure 10. The attenuation and velocity of sound as a function of temperature at a frequency of 20.24 MHz: (c) 23.17 bar.

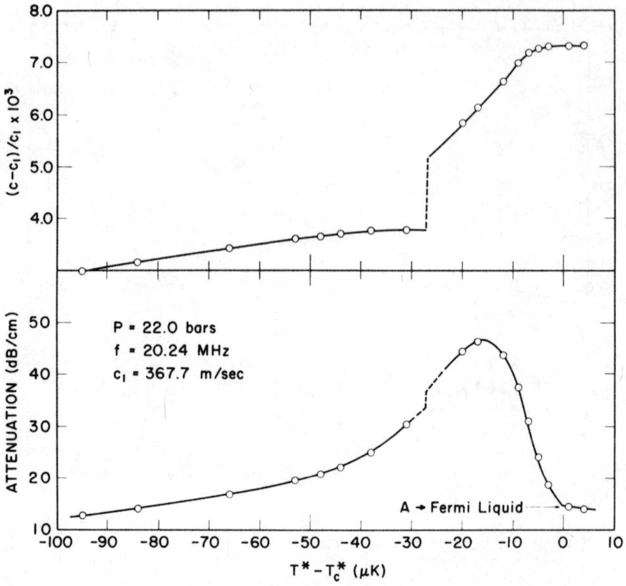

Figure 10. The attenuation and velocity of sound as a function of temperature at a frequency of 20.24 MHz: (d) 22.00 bar.

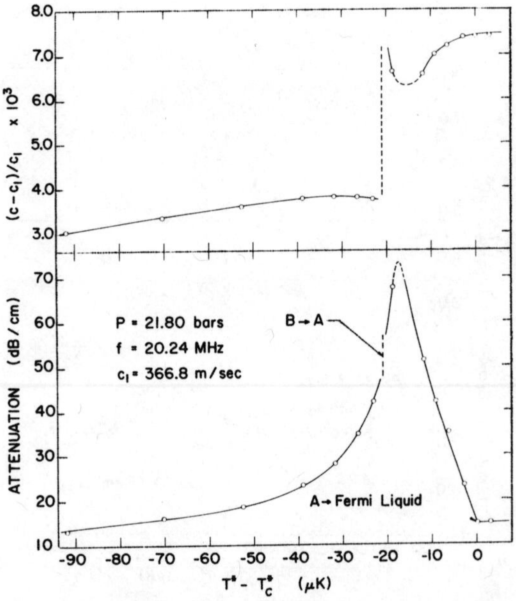

Figure 10. The attenuation and velocity of sound as a function of temperature at a frequency of 20.24 MHz: (e) 21.80 bar.

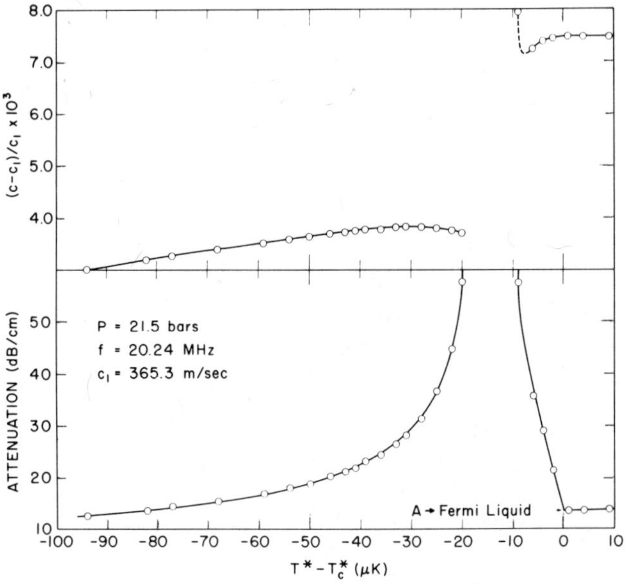

Figure 10. The attenuation and velocity of sound as a function of temperature at a frequency of 20.24 MHz: (f) 21.50 bar.

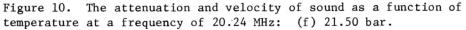

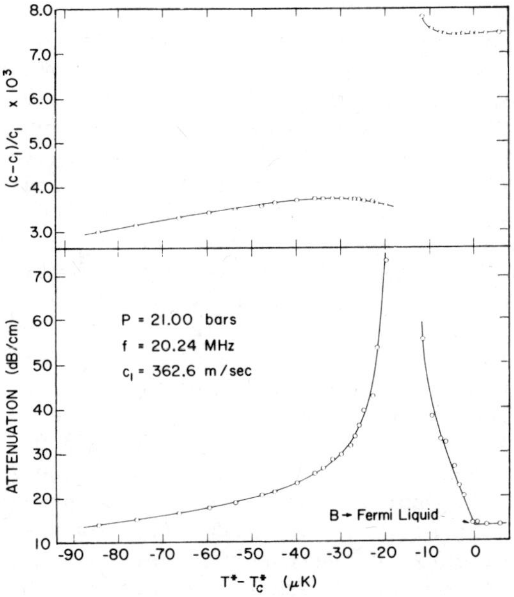

Figure 10. The attenuation and velocity of sound as a function of temperature at a frequency of 20.24 MHz: (g) 21.00 bar.

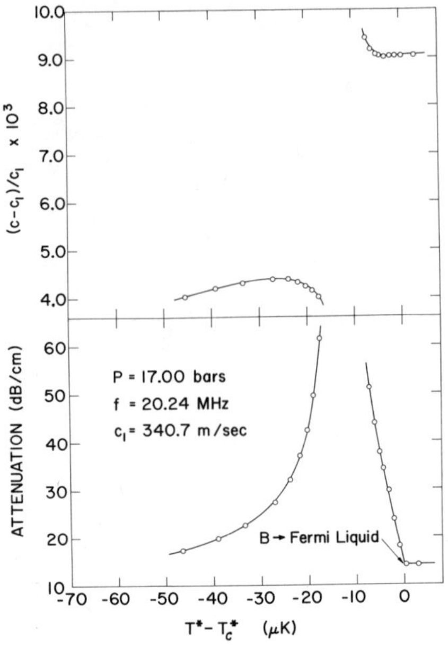

Figure 10. The attenuation and velocity of sound as a function of temperature at a frequency of 20.24 MHz: (h) 17.00 bar.

uncertainty occurring in the normal liquid in the zero sound-to-first sound transition region was settled by comparison with the measurements of Paulson, et al.[38] The normalization of the attenuation data was arrived at by observing the attenuator setting (at each pressure studied) at 450 mK and correcting for the attenuation in the liquid (which is quite low at this temperature) from the viscosity measurements of Betts, et al;[35] fixing the normalization at one temperature yields a nearly T^2 dependence of the zero sound attenuation in the normal liquid. We have identified the appearance of superfluidity as the temperature where the attenuation departs from the normal liquid value; this is in agreement with theory for both phases. The data at higher pressures are similar to those observed by Paulson, Johnson, and Wheatley.[38] Note that a discontinuity in the velocity and attenuation is observed at the B-A transition for most A phase pressures studied (at 28.00 bar the temperature did not extend low enough); this transition was not observed by Paulson, et al., for the pressures and temperatures studied by them. The data in Fig. 10e at 21.80 bar show the situation at a pressure only slightly above the PCP where the A and B transitions are now quite close to each other. Note, both the velocity and attenuation remain discontinuous at the B-A transition (it is still a first order transition) but at temperatures just above and just below the discontinuity the velocity turns up and down, respectively (appearing to anticipate the transition); the velocity still turns down just below T_c, however. The signal was observed continuously through the maximum in the attenuation ($\sim$ 18 μ K below T_c) but at a level near the limit of our signal to noise which was $\sim$ 73 dB/cm, and the peak in Fig. 10e is drawn through this value. In Fig. 10f we show the data at 21.50 bar. The discontinuity associated with the B-A transition is not observed here and presumably occurs near the attenuation maximum where the signal is lost. The velocity is still observed to turn down just below T_c. Figure 10g shows the data at a pressure of 21.00 bar. The "A-like" behavior where the velocity turns down below T_c is now absent. Pressures at which "A-like" behavior is not observed are taken to be below the PCP. Note that the velocity turns down just below the attenuation peak and when it reappears it has the opposite temperature dependence; the behavior near the peak in the attenuation was not observable but the curves are presumably continuous since there is no phase transition at this point. The peculiar structure in the velocity was not observed in the data of Paulson, et al.,[38] although a sharp drop in the velocity at the temperature of the attenuation maximum was seen in their data at 19.6 bar. The data at a pressure well below the PCP (17.00 bar) are shown in Fig. 10h. They are essentially similar to those at 21 bar.

Experiments in a magnetic field were performed in the A phase at 26.0 bar[38b] and in the B phase below the polycritical point at 21.00 bar. Data were collected during temperature drifts for specific fields and angles or during angular rotations at fixed fields and nearly constant temperature; the majority of the data were taken with the first technique. Demagnetizations were performed at the angles ψ = 0, 30, 60 and 90°. At each angle, the fields H = 0, 9, 18, and 27 G were studied.

Figure 11 shows the temperature dependence of the attenuation and velocity shift in the A phase at 26.0 bar for the angles ψ = 0, 30, 60 and 90° and for a field of 27 G. To simplify comparison with theory, the data have been normalized to be zero in the normal liquid. The anisotropy was quite apparent at 9 G and essentially saturated at 18 G. It is quite striking that such small magnetic fields will align the liquid and this observation would appear to open the way for a host of anisotropic transport measurements in the A liquid using CMN refrigeration techniques. Figure 12 shows the angular dependence of the attenuation and velocity shift in the A phase at 26.0 bar in a field of 18 G and at a temperature of 48 μK below T_c. These data have not been normalized to be zero in the normal liquid. The curves through the data are fits to Eq. (5a) and (5b); clearly, the data exhibit an angular dependence that is very close to the predicted form. Recent measurements by Lawson, et al.[38c] on the sound attenuation in the A phase at 20 MHz have also shown anisotropy.

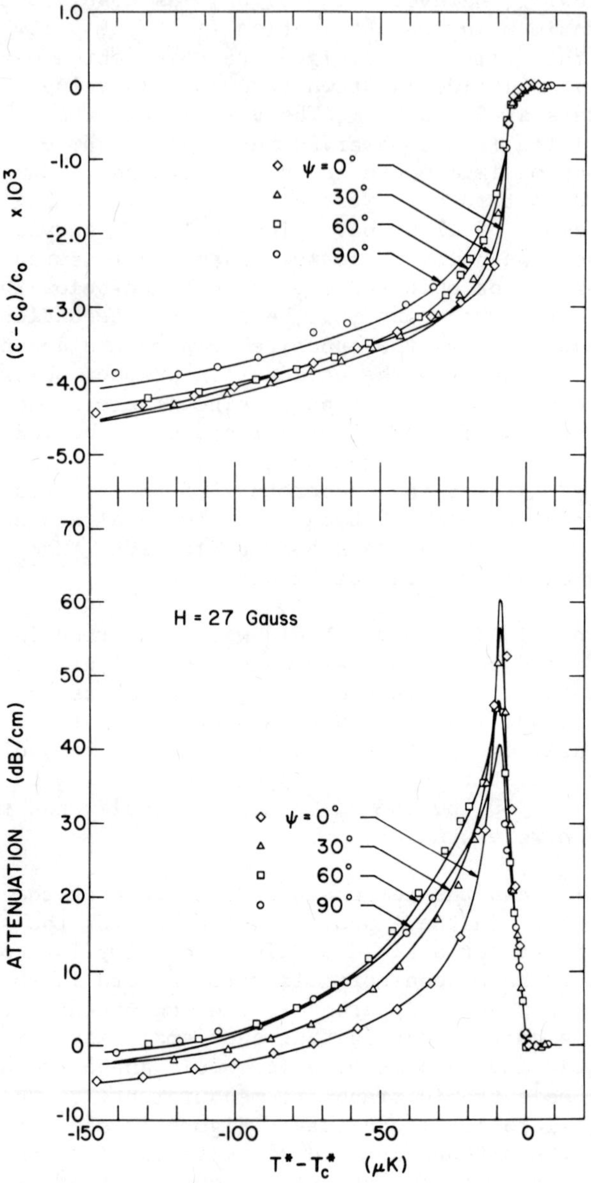

Figure 11. The temperature dependence of the attenuation and velocity shift of 20.24 MHz ultrasound in superfluid ^{3}He A at a pressure of 26.0 bar and for angles, ψ, between the magnetic field and propagation direction of 0°, 30°, 60°, and 90°.

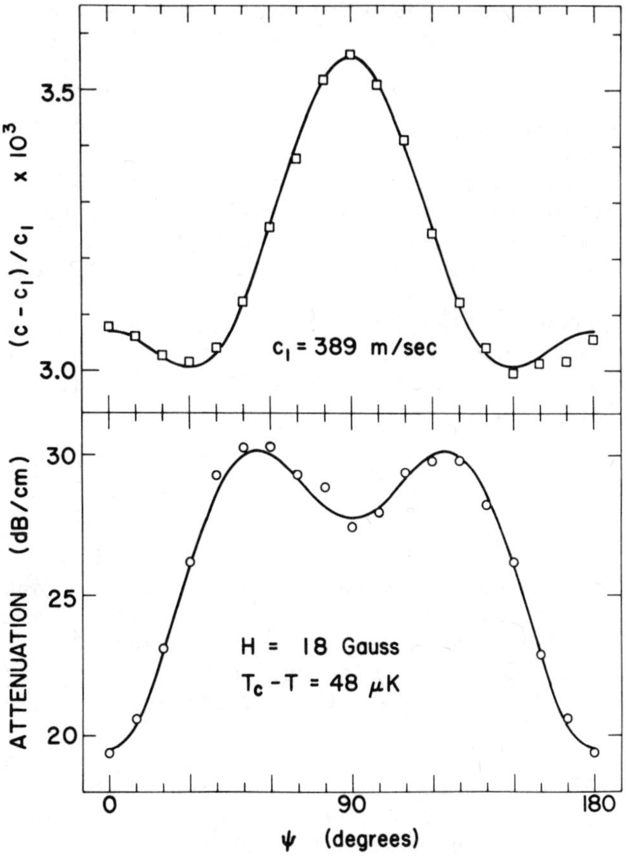

Figure 12. The angular dependence of the attenuation and velocity shift of 20.24 MHz ultrasound in superfluid ^{3}He A for a pressure of 26.0 bar and at a temperature of 48 μK below T_c. The curves are fits of the data to Eqs. (5a) and (5b).

IV. DISCUSSION

Our sound propagation data in the normal fluid present no unusual or unexpected features. The attenuation data in the unoriented superfluid A phase are only in qualitative agreement with the angularly averaged theory of Wölfle[17] and Serene.[20] The width of the main peak is considerably broader than predicted and the small broad peak below the main one (arising from α_c) is not observed. Lifetime effects might account for the widening of the main peak; however, one would not expect the elimination of the second peak. The small sharp feature just above the main peak should not exist according to Wölfle.[20a] Our data on the B phase are in qualitative agreement with features (2), (3), and (4) discussed in Sec. I in connection with Eq. (6), if the effect of lifetime broadening is included; feature (1) could easily be obscured by collision broadening of the δ function peak. Since we were not able to follow the signal through the peak the strength of the δ function (area under the peak) cannot be reliably estimated. However, if we assume that the B phase theory is essentially correct we can use features (3) and (4) to evaluate $\Delta(T)$ at two temperatures; i.e., when $\hbar\omega = 2\Delta(T)$ and $\hbar\omega = 2\sqrt{3/5}\,\Delta(T)$. For the data at 21.00 bar we find $\Delta(T) = 6.6 \times 10^{-20}$ erg and 8.5×10^{-20} erg for $T_C - T = 10.5$ μK and 16.5 μK, respectively. Either of these numbers may be used to compute the specific heat discontinuity, ΔC, at T_C using Eq. (F-13) of Serene[20]

$$\Delta(T)^2 = \frac{2}{3}\pi^2(k_B T_C)\frac{\Delta C}{C_N}\frac{(T_C - T)}{T_C}$$

where C_N is the normal state heat capacity at the transition; we find $\Delta C/C_N = 1.5$ which is consistent with a low pressure extrapolation of the data of Webb, et al.,[39] who found $\Delta C/C_N = 1.85$ at 34 bars and 1.55 at 23 bars.[40] The agreement is impressive and lends support to the identification of the B phase as a Balian-Werthamer state.[16a,15]

The effect of quasiparticle scattering has been studied recently by Wölfle[20a] and Maki and Ebisawa.[20b] Wölfle's calculations for the B phase are in excellent agreement with the data of Paulson, et al,[38] for both the velocity and attenuation at 15.15 MHz and 19.6 bar. However, his calculation does not show the structure in the velocity that we observe at 20.24 MHz. The calculation by Maki and Ebisawa,[20b] on the other hand, shows striking structure in the velocity that is quite similar to the structure that is suggested by our data. Their calculation also predicts a low-temperature limit to the sound velocity that is slightly different from the hydrodynamic first sound velocity c_1.

We now discuss our attenuation and velocity measurements in the presence of a magnetic field. We have used the data of Fig. 11 to deduce the temperature dependence of each of the combination of coefficients entering Eq. (5). From these combinations we have solved for the temperature dependent coefficients $\alpha_{||}(T)$, $\alpha_c(T)$, $\alpha_\perp(T)$, $\Delta c_{||}(T)$, $\Delta c_c(T)$, and $\Delta c_\perp(T)$. The temperature dependence of these coefficients is shown in Fig. 13. The dashed lines show calculations using the programs of Serene for this pressure and frequency. Qualitatively the reduced data and the calculations for the attenuation are quite similar; $\alpha_\perp$ has a sharp peak, α_c has a

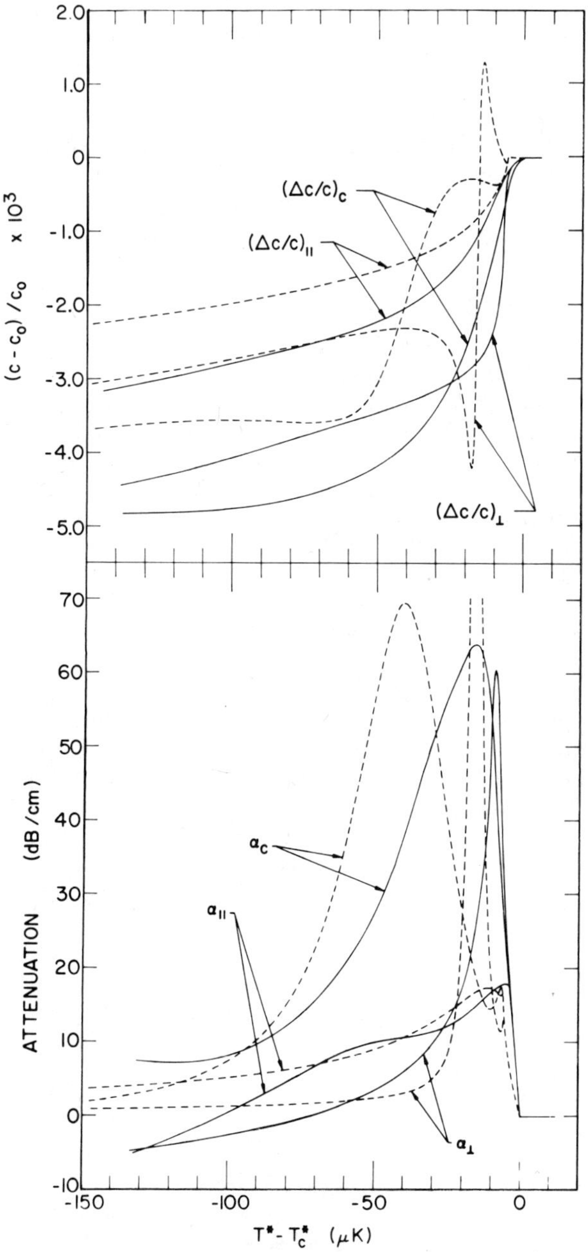

Figure 13. The temperature dependence of the quantities which determine the anisotropic attenuation and velocity shift for 20.24 MHz sound in ^{3}He A at 26.0 bar; the dashed lines are the calculations of Serene (Ref. 20).

broad peak at a lower temperature than that of $\alpha_\perp$, and $\alpha_{||}$ rises quickly to a small value and then falls slowly. The theory does not include quasiparticle lifetime effects and the small sharp structure just below T_c predicted by the theory could easily be obscured or, as pointed out by Wölfle,[20a] be an artifact of the computer calculations. The fact that experimentally the peaks in $\alpha_\perp$ and α_c are occurring at higher temperatures suggests that the temperature dependence of the gap is larger than estimated by Serene from the heat capacity anomaly on entering the A phase; this would explain the problem raised in the beginning of this section. The agreement between theory and experiment for the velocity anisotropy is less satisfying; here lifetime effects may play a dominant role.

We have also made a search for anisotropy in the sound propagation in the B phase of superfluid ^{3}He. Our measurements were made at a pressure of 21.0 bar which is slightly below the polycritical point. No anisotropy was observable within our resolution. This observation supports the conjecture that the B phase is a Balian-Werthamer[16a] state for which an isotropic gap (and thus no anisotropy) is predicted.

ACKNOWLEDGMENTS

In conclusion we would like to thank K. Maki for preprints and helpful correspondence on the theory, R. Roger for aid in fabricating the mechanical heat switch, J. W. Serene for sending us his program for the A phase anisotropy calculations, P. Wölfle, and H. Ebisawa and K. Maki for sending us preprints of their latest work which includes quasiparticle scattering, and M. Kuchnir for aid with the early phases of the experiments.

References

(1) L. D. Landau, Zh. Eksp. Teor. Fiz. 32, 59 (1957), [Sov. Phys.-JETP 5, 101 (1957)].

(2) W. R. Abel, A. C. Anderson, and J. C. Wheatley, Phys. Rev. Lett. 17, 74 (1966).

(3) C. J. Pethick, Phys. Rev. 185, 384 (1969).

(4) G. Baym and C. Pethick, "Landau Fermi Liquid Theory and Low Temperature Properties of Liquid ^{3}He," in The Physics of Liquid and Solid Helium, Vol. II, edited by K. H. Bennemann and J. B. Ketterson (John Wiley, N.Y.) to be published.

(5) L. D. Landau and E. M. Lifshitz, Fluid Mechanics (Pergamon Press, Oxford, 1959).

(6) A. A. Abrikosov and I. M. Khalatnikov, Rept. Progr. Phys. 22, 329 (1959).

(7) G. A. Brooker, Proc. Roy. Soc. (London) 90, 397 (1967).

(8) D. D. Osheroff, R. C. Richardson, and D. M. Lee, Phys. Rev. Lett. 28, 885 (1972).

(9) T. A. Alvesalo, Yu. D. Anufriyev, H. K. Collan, O. V. Lounasmaa, and P. Wennerstrom, Phys. Rev. Lett. 30, 962 (1973).

(10) H. J. Kojima, D. N. Paulson, J. C. Wheatley, Phys. Rev. 32, 141 (1974).

(11) T. J. Greytak, R. T. Johnson, D. N. Paulson, and J. C. Wheatley, Phys. Rev. Lett. 31, 452 (1973).

(12) W. J. Gully, D. D. Osheroff, D. T. Lawson, R. C. Richardson, and D. M. Lee, Phys. Rev. A 8, 1633 (1973).

(13) A. I. Ahonen, M. T. Haikala, M. Krusius, and O. V. Lounasmaa, Phys. Rev. Lett. 33, 628 (1974).

(14) P. W. Anderson and P. Morel, Phys. Rev. 123, 1911 (1961).

(15) P. W. Anderson and W. Brinkman, Phys. Rev. Lett. 30, 1108 (1973).

(16) A. J. Leggett, Phys. Rev. Lett. 31, 352 (1973).

(16a) R. Balian and N. R. Werthamer, Phys. Rev. 131, 1553 (1963).

(17) P. Wölfle, Phys. Rev. Lett. 30, 1169 (1973); ibid. 1437 (1973).

(18) H. Ebisawa and K. Maki, Progr. Theor. Phys. 51, 337 (1974).

(19) K. Maki, J. Low Temp. Phys. 16, 465 (1974), and private communication.

(20) J. W. Serene, "Theory of Collisionless Sound in Superfluid ^{3}He," Thesis, Cornell University (1974) and to be published.

(20a) P. Wölfle, these proceedings, and to be published.

(20b) K. Maki and H. Ebisawa, to be published.

(21) We thank Dr. Serene for sending us his program for computing these quantities.

(22) K. Miyano and J. B. Ketterson, Phys. Rev. Lett. 31, 1047 (1973).

(23) K. Miyano and J. B. Ketterson, Phys. Rev. to be published.

(24) B. M. Abraham, Y. Eckstein, J. B. Ketterson, and J. H. Vignos, Cryogenics 9, 274 (1969).

(25) P. R. Roach, J. B. Ketterson, B. M. Abraham, J. Monson, and P. D. Roach, Rev. Sci. Instr. 46, 207 (1975).

(26) Paul D. Roach, B. M. Abraham, P. R. Roach, and J. B. Ketterson, to be published.

(27) B. M. Abraham, D. W. Osborne, and B. Weinstock, Phys. Rev. 75, 988 (1949).

(28) K. N. Zinovieva, Zh. Eksp. Teor. Fiz. 34, 609 (1958); [Sov. Phys.-JETP 7, 421 (1958)].

(29) R. D. Taylor and J. G. Dash, Phys. Rev. 132, 2372 (1965).

(30) D. S. Betts, D. W. Osborne, B. Welber, and J. Wilks, Phil. Mag. 8, 977 (1963).

(31) R. W. H. Webeler and D. C. Hammer, Phys. Lett. 21, 403 (1966).

(32) M. A. Black, H. E. Hall, and K. Thompson, J. Phys. C 4, 129 (1971).

(33) W. R. Abel, A. C. Anderson, and J. C. Wheatley, Phys. Rev. Lett. 7, 299 (1961); Phys. Rev. Lett. 17, 74 (1966).

(34) B. M. Abraham, D. Chung, Y. Eckstein, J. B. Ketterson, and P. R. Roach, J. Low Temp. Phys. 6, 521 (1972).

(35) D. S. Betts, B. E. Keen, and J. Wilks, Proc. Roy. Soc. A 298, 34 (1965).

(36) V. J. Emery, Phys. Rev. 175, 251 (1968).

(37) D. T. Lawson, W. J. Gully, S. Goldstein, R. C. Richardson, and D. M. Lee, Phys. Rev. Lett. 30, 541 (1973).

(38) D. N. Paulson, R. T. Johnson, J. C. Wheatley, Phys. Rev. Lett. 30, 829 (1973).

(38a) Pat R. Roach, B. M. Abraham, M. Kuchnir, and J. B. Ketterson, Phys. Rev. Lett. 34, 711 (1975).

(38b) Pat R. Roach, B. M. Abraham, P. D. Roach, and J. B. Ketterson, Phys. Rev. Lett. 34, 715 (1975).

(38c) D. T. Lawson, H. M. Bozler, and D. M. Lee, Phys. Rev. Lett. 34, 121 (1975).

(39) R. A. Webb, T. J. Greytak, R. T. Johnson, and J. C. Wheatley, Phys. Rev. Lett. 30, 210 (1973).

(40) J. C. Wheatley, Physica 69, 218 (1973).

PHONON PROPAGATION IN LIQUID AND SOLID HELIUM

V. Narayanamurti and R. C. Dynes

Bell Laboratories

Murray Hill, N. J. 07974

Abstract: A brief review is given of recent work on the measurement of anomalous dispersion in the excitation spectrum of Helium II, using superconducting fluorescers and tunnel junctions as sources and detectors of high frequency phonons. In addition, a short discussion of recent observations of ballistic phonons in solid ^{3}He is given. Implications for specific heat measurements are discussed.

I. INTRODUCTION

The question of phonon scattering in liquid He II below 0.6 K has been a subject of active interest in recent years: This subject has been proved primarily through ultrasonic[1] and thermodynamic[2] measurements. In this paper we summarize the results of our work on high frequency "monochromatic" phonon propagation which yields information on the phonon scattering mechanism. We show that these data provide strong evidence for the 3 phonon process (3 pp) being the dominant one. Consistent with this, the data shows that the excitation spectrum in He II ($\omega(k)$) curves upward at low momenta (anomalous dispersion). A detailed account of this work is published elsewhere.[3]

We also give a brief account of very recent work on heat pulse propagation in solid ^{3}He for crystals with molar volume ~ 24 cm^3/mole. Observations of ballistic phonon propagation, coupled with the temperature dependence of the phonon-phonon scattering are described. The implications for recent specific heat measurements[4] are discussed. Again, a detailed account will be published elsewhere.[5]

II. EXPERIMENTAL RESULTS AND DISCUSSION

A. Liquid Helium II

From earlier work in solids,[6] it is known that a superconducting tunnel junction emits a discrete spectrum of high frequency phonons ($\hbar\omega >> kT$) which can be altered by varying the applied voltage to the generating junction or by the application of a magnetic field. Utilizing this and a tunnel junction detector we have determined that phonons below a cutoff energy E_c are subject to spontaneous decay via the 3 phonon process (3 pp). Above E_c the phonons are shown to be long lived and propagate macroscopic distances. This cutoff energy varies with pressures and the dependence is shown as the circles in Fig. 1.

We see that E_c versus pressure is a smooth function which apparently goes to zero in the vicinity of 20 bars. Within small corrections, one can convert E_c to a wave number cutoff k_c and this is shown as the squares in Fig. 1. Here we see an approximately linear function of pressure and at SVP, k_c equals $.51 \pm .02\text{Å}^{-1}$. This implies that phonons of wavelength less than 12Å are sufficiently long lived to propagate macroscopic distances while those of long wavelength are scattered. For a model excitation spectrum of the form

$$\omega(k) = C_o k(1-\gamma k^2-\delta k^4....) \tag{1}$$

where C_o is the phonon velocity (sound velocity) at $k = 0$, and γ and δ are dispersion parameters, the cutoff momentum k_c is given by

$$k_c^2 = -\frac{4}{5}\frac{\gamma}{\delta} \tag{2}$$

and the group velocity $\frac{d\omega}{d_k}$ at this spontaneous decay cutoff is given by

$$v_g(k_c) = C_o(1+\gamma k_c^2) \tag{3}$$

Thus for $\gamma < 0$, $v_g(k_c) < C_o$. From the measured values of k_c and $v_g(k_c)$, one can obtain estimates for γ and δ. In addition to the measurements of k_c shown in Fig. (1), $v_g(k_c)$ has been experimentally determined.[3] At SVP for example $v_g(k_c) = 231 \pm 1.5$ m/sec compared with a C_o of 238.3 m/sec. We regard this as evidence for anomalous dispersion below k_c and from this measure we estimate γ and δ to be respectively $.12 \pm .025\ \text{Å}^2$ and $0.36 \pm 0.09\ \text{Å}^4$. Note that the values of γ and δ are highly model dependent. For a detailed comparison with other models the reader is referred to Ref. 3.

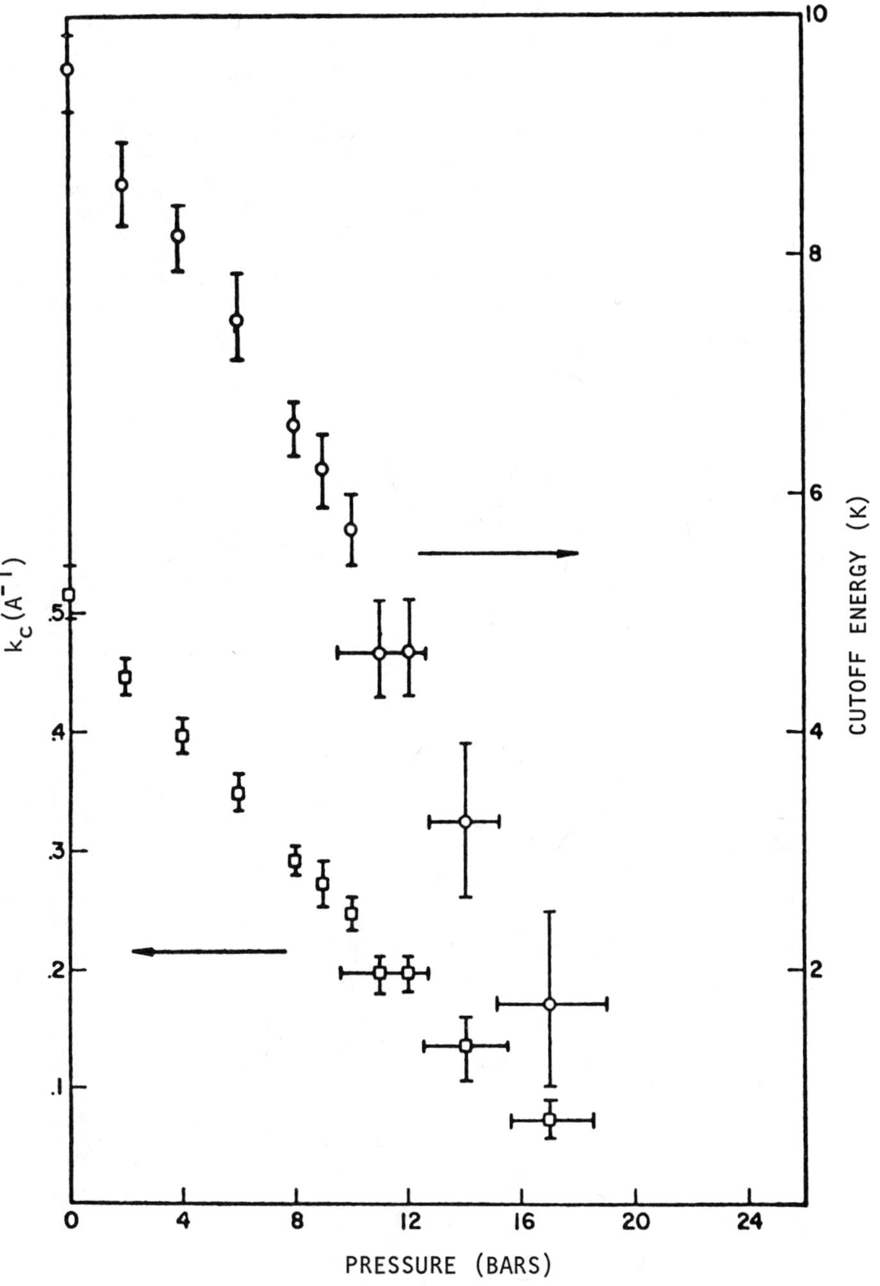

Figure 1. E_c (circles) and calculated k_c (squares) as a function of pressure. Liquid He II.

B. Solid Helium

In a pure dielectric solid, depending on the temperature, heat can propagate without interaction (ballistic phonons), as a true temperature wave (collective second sound), or by diffusion. In this section we give a very brief description of our observations of the transition from second sound to ballistic phonon flow in solid ^{3}He.

In Fig. 2 we have plotted the temperature dependence of the heat pulse velocities for a crystal of solid ^{3}He with a molar volume

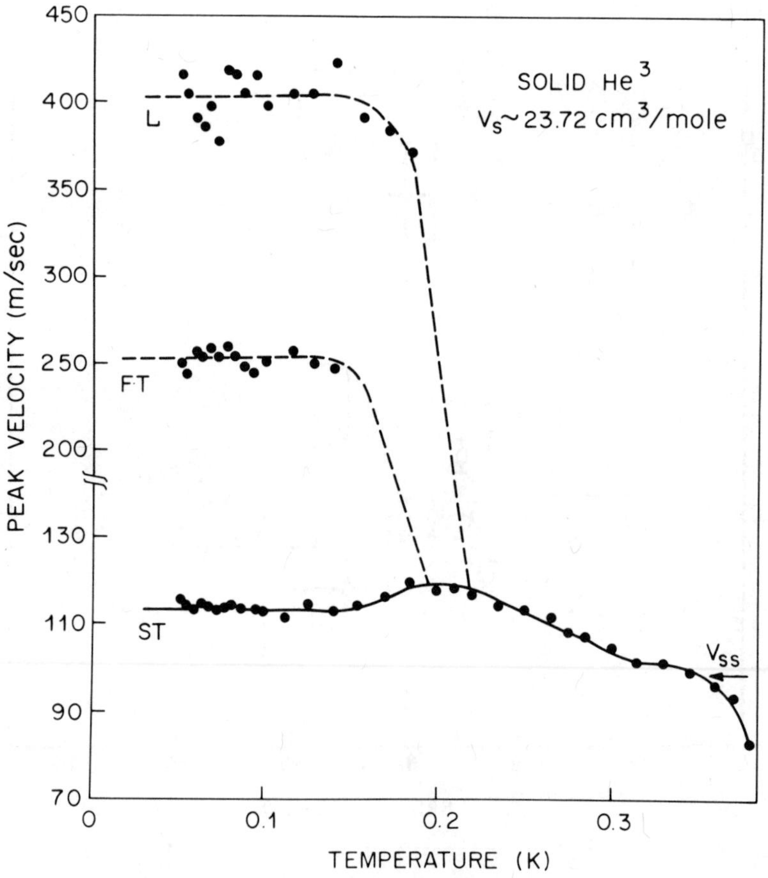

Figure 2. Peak heat pulse velocities as a function of temperature in solid ^{3}He. Molar volume 23.72 cm^3/mole. Orientation [110].

of 23.72 cm^3/mole. At low temperatures ($\lesssim$ .1 K) the limiting velocities of the longitudinal (L), fast transverse (FT), and slow transverse (ST) modes indicate an orientation close to [110] by comparison with Greywall's sound velocity data.[7] As the temperature is raised to 0.2 K the ST mode velocity increases from 113 m/sec to 119.5 m/sec. It then decreases to a value of 101 m/sec in the temperature range 0.3 to 0.34 K. In this region the propagating mode is second sound. At higher temperatures diffusion is observed.

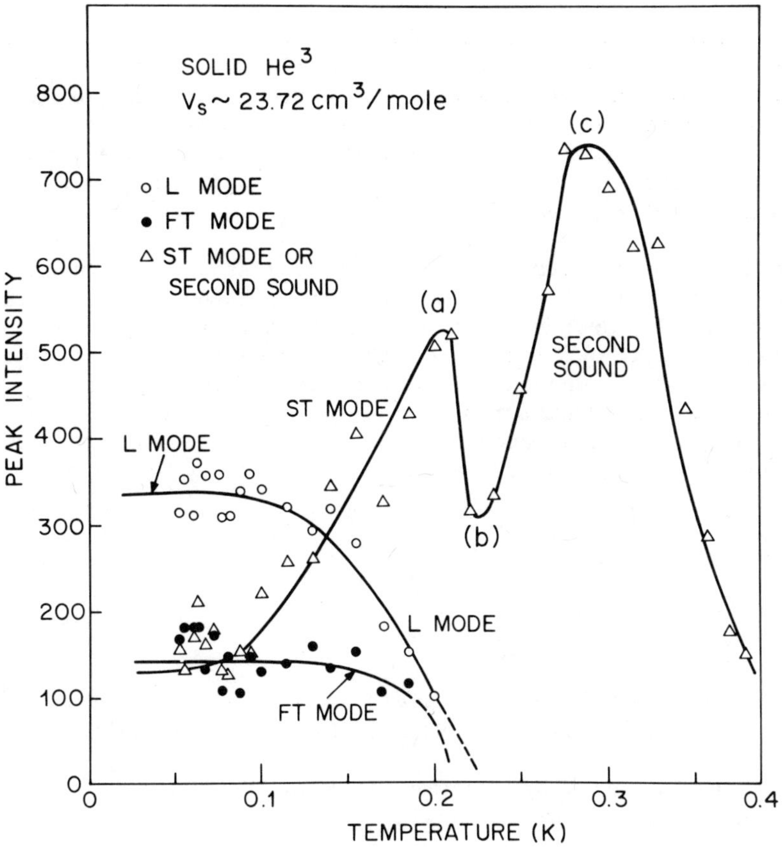

Figure 3. Peak intensity of heat pulses in solid He as a function of temperature. Molar volume 23.72 cm^3/mole.

The velocity behavior for the [110] ST phonon described above was also observed in a crystal of molar volume 24.45 cm^3/mole. No unusual behavior was observed with crystals possessing a [100] orientation or in similar measurements on solid ^4He. A clue to this unusual behavior is obtained by plotting the intensities of the three pulses (Fig. 3). In the region between 0.15 to 0.2 K, where the L and FT modes rapidly decay, the ST mode picks up in intensity. Above 0.22 K, a rapid decay of the ST mode intensity also occurs consistent with observed velocity dispersion. A second maximum is reached at about 0.3 K as the collective second sound mode forms.

The velocity data of Fig. 2 are, therefore, interpreted as a mode pulling effect due to the higher probability for decay of the L and FT branches compared to the ST branch[5] by the 3 phonon process. At higher temperatures, the ST mode eventually also decays via higher order anharmonic N-processes and one observes second sound.

This interpretation is also consistent with calculations of the expected second sound velocity in an elastically anisotropic medium. Using Kwok's formula[8] and Greywall's elastic constants[7] we have calculated the second sound velocity to be 96.3 m/sec for a molar volume of 23.72 cm /mole. This is very close to the measured velocity of 101 m/sec. The elastic Debye velocity divided by $\sqrt{3}$ has on the other hand a value of 135 m/sec.

Finally, the maximum in the ST velocity as a function of temperature implies that the density of states and hence the lattice specific heat, C_V, must show an unusual temperature dependence.[5] Note, however, that our data imply that the lattice specific heat (due to the phonons) should show the usual T^3 behavior below .12 K. Thus, the observed[4] departures from T^3 below .12 K must be due to other causes, while that above could in part be due to the unusual temperature dependence of the slow mode.

References

(1) P. R. Roach, J. B. Ketterson and M. Kuchnir, Phys. Rev. A5, 2205 (1972).

(2) N. E. Phillips, C. A. Waterfield and J. K. Hoffer, Phys. Rev. Lett. 25, 1260 (1970); H. Maris, Phys. Rev. A8, 1280 (1973).

(3) R. C. Dynes and V. Narayanamurti, Phys. Rev. Lett. 33, 1195 (1974) and to appear in Physical Review.

(4) S. H. Castles and E. D. Adams, Phys. Rev. Lett. 30, 1125 (1973).

(5) V. Narayanamurti and R. C. Dynes, to appear in Physical Review.

(6) R. C. Dynes and V. Narayanamurti, Phys. Rev. B6, 143 (1972) and references cited therein.

(7) D. S. Greywall, Phys. Rev. B11, 1070 (1975).

(8) P. C. Kwok, Physics 3, 221 (1967).

THERMAL BOUNDARY RESISTANCE BETWEEN SOLID ^{3}He AND CERIUM MAGNESIUM NITRATE

L. E. Reinstein and George O. Zimmerman

Department of Physics, Boston University
Boston, Ma. 02215

and Francis Bitter National Magnet Laboratory*
M.I.T., Cambridge, Ma. 02139

We have measured the thermal boundary resistance between powdered cerium magnesium nitrate (CMN), (average particle diameter ∿50μ) and solid He^3 (23.9 cm^3/mole, ∿15ppm He^4) and liquid He^3 (S.V.P.) at several different applied magnetic field strengths in the temperature range between 45 mK and 250 mK. At temperatures below 70 mK and fields greater than ∿55G, the observed measurements of the magnetic Kapitza resistance are consistent with the T^2 dependence of that resistance predicted by R. A. Guyer[1] for the solid He^3 and the T dependence for the liquid observed by others at lower temperatures[2,3] and theoretically predicted.[1,4,5]

The method used to measure the boundary resistance was similar to that of Ref. 2. The experimental chamber, shown in Fig. 1 was thermally tied to a dilution refrigerator and contained 0.5 gm of CMN and 2.7 cm^3 of He^3. These quantities were chosen so that the specific heat of the He^3 was always significantly greater than that of CMN. To insure that the CMN was coupled to the He^3 rather than directly to the dilution regrigerator, experiments were also carried out with no He^3 in the chamber, with only He^3 vapor and with liquid He^3. The temperature was measured by means of a resistor calibrated against the CMN susceptibility at zero magnetic field. As an added check, it was determined that at the temperatures of our measurement of the relation between $\chi(0)$, the susceptibility of CMN in zero

*Supported by the National Science Foundation.

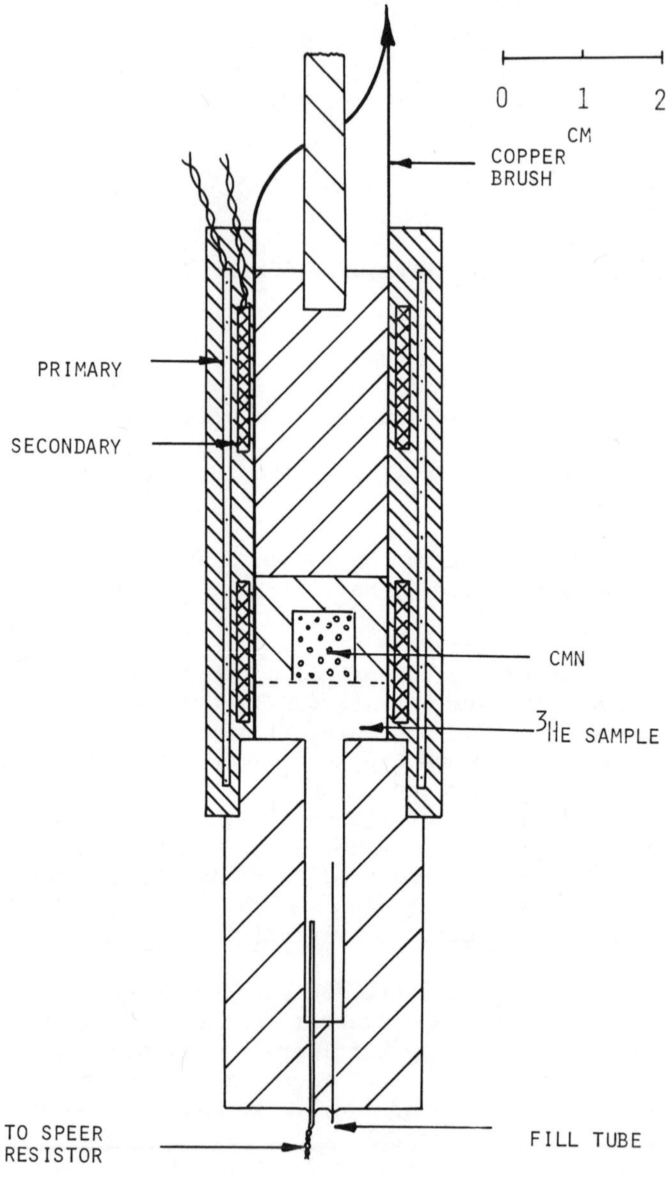

Figure 1

magnetic field, and $\chi(H)$, the susceptibility in a dc magnetic field, H, parallel to the susceptibility measuring field was given by[6]

$$\frac{\chi(0)}{\chi(H)} - 1 \propto H^2 \qquad (1)$$

to a high precision.

By applying a magnetic field and then reducing it to a desired field value, we quickly demagnetized the CMN powder, thereby lowering its temperature ($\frac{\Delta T}{T} \sim 3\%$) and then observed the subsequent equilibrium time constant as it warmed back to the ambient temperature of the He^3. These time constants are shown in Fig. 2 for the solid and Fig. 3 for the liquid at different final magnetic fields.

The time constants at H=0, shown in Fig. 4 (H is the applied magnetic field) fit a straight line described by

$$\tau_{P.B.} = 7.4 \times 10^{-3} T^{-2} \text{ sec} \tag{2}$$

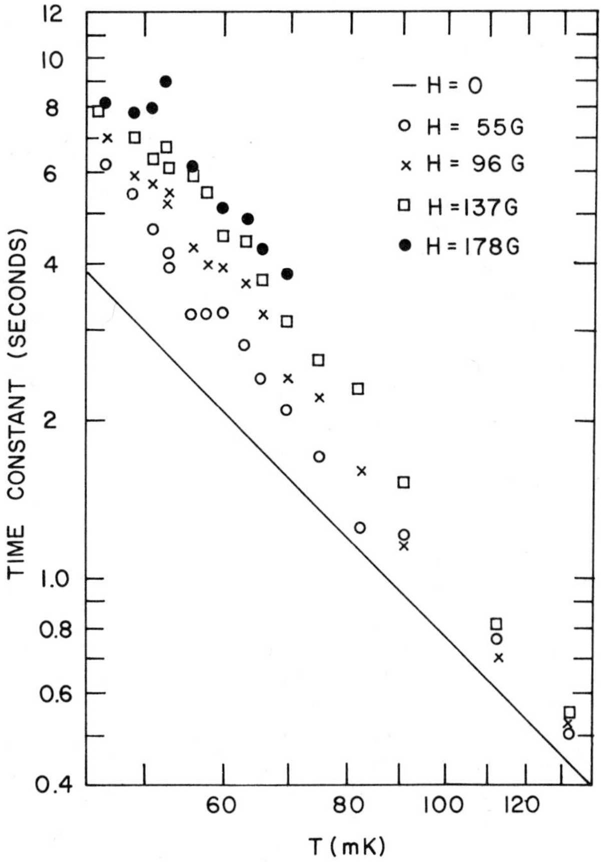

Figure 2

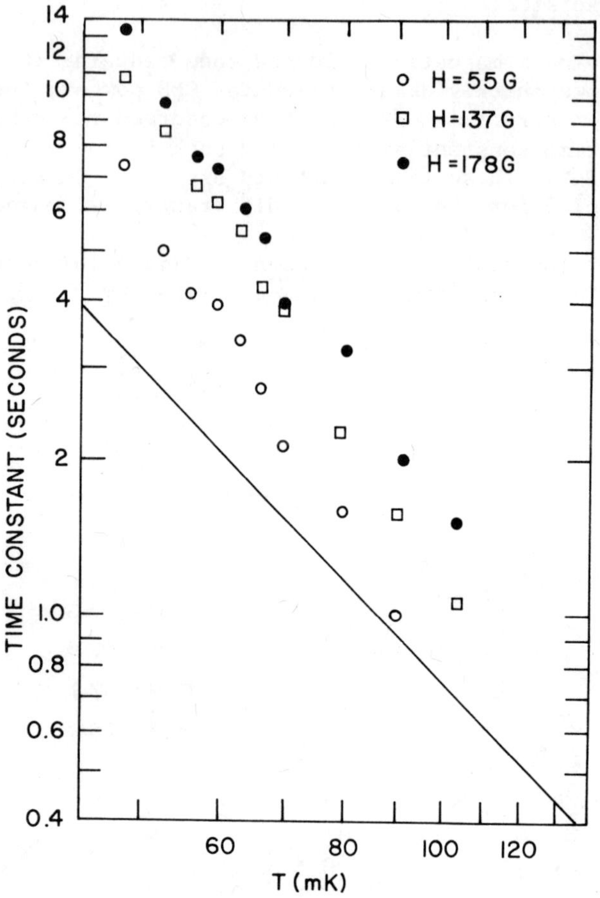

Figure 3

which is the expected "phonon bottleneck" observed in CMN at these temperatures.[7] It has been shown[8] that the phonon bottleneck time constant in CMN, $\tau_{P.B.}$ is field independent in the high temperature approximation. Thus, if one assumes that the phonon bottleneck is in series with the total thermal boundary resistivity ρ_B, which is associated with the time constant, τ_B, through the relationship

$$\tau_B = (V/A)\ c_V(H)\rho_B \tag{3}$$

where $c_V(H)$ is the heat capacity of CMN per unit volume[9] and equals

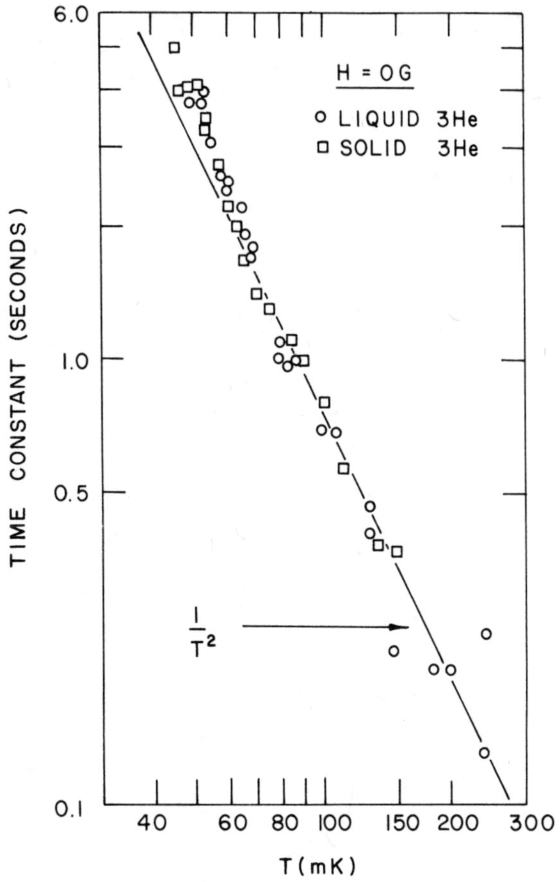

Figure 4

$c_V(0) + CH^2/T^2$ where C is the Curie constant, V is the CMN volume and A is its surface area,[10] $\tau_B(H,T)$ can be determined by measuring the total relaxation time, $\tau(H,T)$ in a magnetic field at a temperature T, and subtracting the value of the field independent $\tau_{P.B.}(T)$, given by Eq. (2). These results for ρ_B are shown for the solid in Fig. 5 and the liquid in Fig. 6.

The value of V/A for the powdered CMN sample was estimated according to the method used by Bishop, Cutter, Mota and Wheatley.[3] However, the reliability of this estimate is in question and might possibly be smaller by an order of magnitude because the particles are in contact with each other and have anisotropic heat conductivity. Thus the absolute value of ρ_B is only approximate and the significant result is the dependence of ρ_B on T.

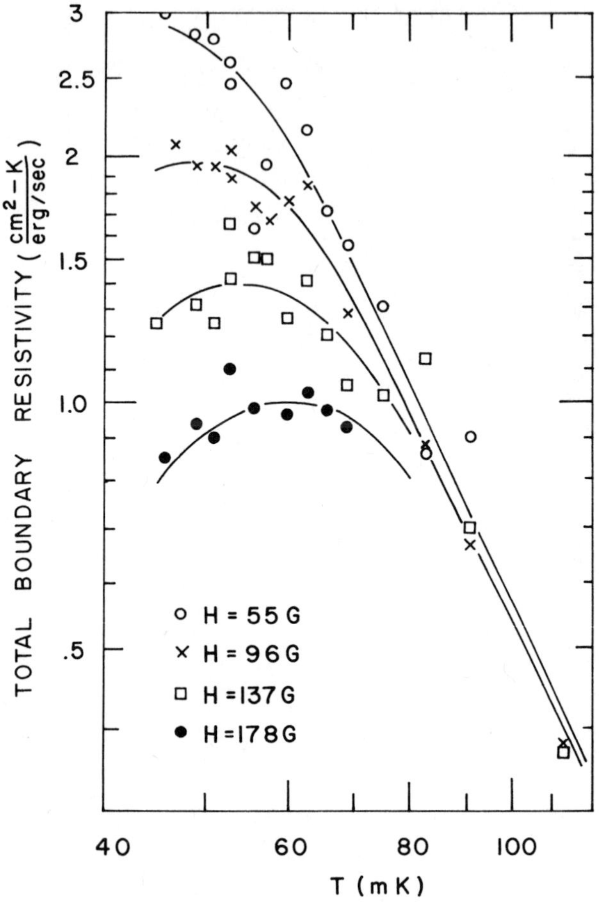

Figure 5

The total thermal boundary resistivity, consists of a parallel combination of phonon resistivity, ρ_P, and magnetic dipolar resistivity, $\rho_m(H)$.[3] We can write

$$\rho_B(H) = \rho_P \rho_m(H) \ / \ (\rho_P + \rho_m(H)) \tag{4}$$

By assuming the field independent ρ_P takes the form $\rho_P \alpha T^{-3}$, we calculated from the data in Fig. 5 and Fig. 6 that

(Solid) $\rho_P = 5.7 \times 10^{-4}((K^4\text{-cm}^2)/(\text{erg/sec}))T^{-3}$.

(Liquid) $\rho_P = 5.5 \times 10^{-4}((K^4\text{-cm}^2)/(\text{erg/sec}))T^{-3}$. (5)

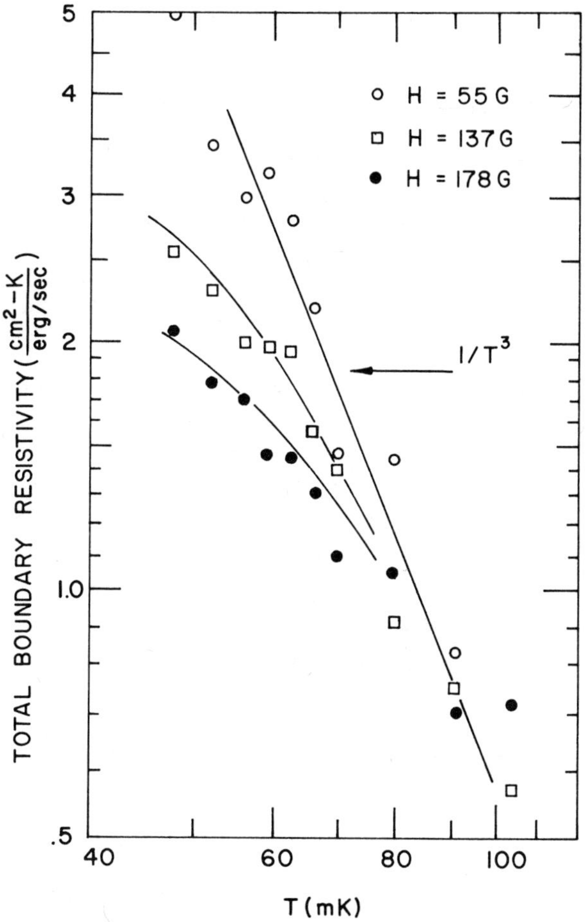

Figure 6

From this, the values of $\rho_B(H,T)$, and Eq. 4 we obtain $\rho_m(H,T)$. In order to compare our data with theoretical prediction, it was useful to relate $\rho_m(H,T)$ to $\rho_m(0,T)$. Using a Redfield-like theory,[11] J. Bishop derives the relation

$$\rho_m(H,T) = (H_d^2/(H_d^2 + \frac{1}{2} H^2))\ \rho_m(0,T) \qquad (6)$$

where H_d is the CMN mean square local dipolar field and H is the applied field. Eq. 6 is valid for the condition of H, $H_d \ll k_B T/\gamma h$ (γ is the CMN gyromagnetic ratio) which holds in our measurements, and implies that one improves thermal contact by increasing the applied field.

If we assume H_d = 40G and apply Eq. 6 to the obtained $\rho_m(H,T)$ values we obtain the data shown in Fig. 7 for the solid. It is seen here that our data are in agreement with the predicted form for the solid[1]

$$\rho_m(0,T) = \text{const.} \times T^2 \quad \text{(solid)} \qquad (7)$$

with a mean square deviation of ∿15%. The constant, (which, as stated before is not reliable) is equal to 5×10^3 cm^2-sec/erg-K.

The data for $\rho_m(0,T)$ in the liquid case are presented in Fig. 8. They show agreement with the equation

$$\rho_m(0,T) = 760 \frac{cm^2}{erg/sec} T \quad \text{(liquid)} \qquad (8)$$

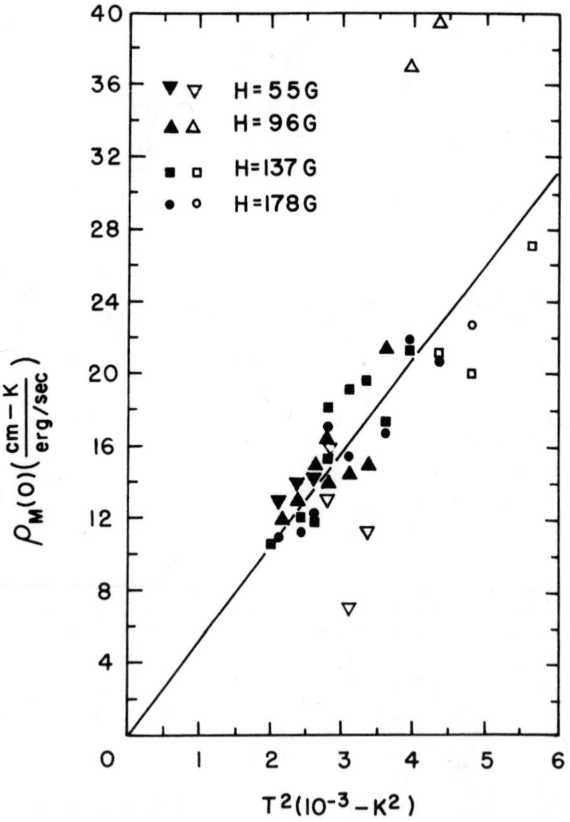

Figure 7

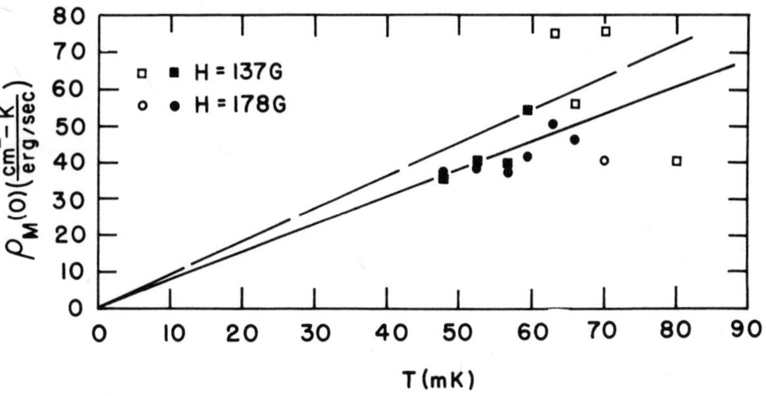

Figure 8

with a 20% mean square deviation in the temperature region between 50 mK and 70 mK. For comparison the coefficient of T in Eq. 8 is given as 900 in (BCMW)[3] and as 1400 in (BMWBB).[3] The precision of the liquid measurement was smaller than that of the solid because at the lowest temperatures, about 50 mK, $\rho_m(0,T)$ of the liquid is a factor of 5 greater than that of the solid. The observations of $\rho_m(0,T)$ in the liquid at these temperatures seems to contradict the interpretation by Harrison and Pendrys[12] of the data of Bishop et al. in terms of only phonon Kapitza boundary resistance and the phonon bottleneck.

To our knowledge, this is the first experimental observation of the T^2 dependence of ρ_m between CMN and solid He^3. Although the precision of the data is not sufficient to establish the T^2 dependence precisely in the solid He^3 case, the mean square deviation of the data from this law is 15% and thus it definitely rules out the T dependence obeyed between CMN and liquid He^3. This result is also significant because of its implication on paramagnetic cooling of He^3 solid by CMN where the limiting resistance will now become the phonon bottleneck. $\tau_{P.B.}$, however, depends on the particle size[12] and could thus be made small. Although CMN undergoes a paramagnetic to antiferromagnetic transition at 1.6 mK, samples have been cooled down to 0.4 mK[13] by adiabatic demagnetization, and thus CMN should be able to cool solid He^3 to temperatures below 1mK.

References

(1) R. A. Guyer, J. Low Temp. Phys. 10, 157 (1972).

(2) W. R. Abel, A. C. Anderson, W. C. Black, and J. C. Wheatley, Phys. Rev. Lett. 16, 273 (1966).

(3) W. C. Black. S. C. Mota, J. C. Wheatley, J. H. Bishop and P. M. Brewster, J. Low Temp. Phys. 4, 391 (1971); also see J. H. Bishop, D. W. Cutter, A. C. Mota and J. C. Wheatley, J. Low Temp. Phys. 10, 379 (1973).

(4) A. J. Leggett and M. Vuorio, J. Low Temp. Phys. 3, 359 (1970).

(5) D. L. Mills and M. T. Beal-Monod, Phys. Rev. A 10, 343 (1974).

(6) B. M. Abraham, J. B. Ketterson and P. R. Roach, Phys. Rev. B 6, 4675 (1972).

(7) A. C. Anderson and J. E. Robichauz, Phys. Rev. B 3, 1410 (1971).

(8) K. W. Mess, thesis, Leiden, 1969 (unpublished).

(9) K. W. Mess, J. Lubbers, L. Nielsen, W. J. Huiskamp, Physica 41, 260 (1969).

(10) Eq. 3 assumes that the CMN is in thermal contact with a reservoir at constant temperature. We have observed this to be true under all the conditions of our experiment.

(11) J. Bishop, thesis, University of California at San Diego, 1973 (unpublished).

(12) J. P. Harrison and J. P. Pendrys, Phys. Rev. B 8, 5940 (1973).

(13) G. O. Zimmerman, E. Maxwell, D. S. Abeshouse and D. R. Kelland, (unpublished).

NUCLEAR SPIN ORDERING OF SOLID ^{3}He IN A MAGNETIC FIELD*

R. B. Kummer, E. D. Adams, W. P. Kirk, A. S. Greenberg,† R. M. Mueller, C. V. Britton, and D. M. Lee‡

Department of Physics and Astronomy, University of Florida

Gainesville, Florida 32611

Abstract: The effect of an external magnetic field on the nuclear spin ordering in solid ^{3}He has been observed for fields up to 1.2 T. For B $\lesssim$ 0.42 T the ordering is similar to the zero field transition reported by Halperin, et al. and the transition temperature is depressed by the field. In higher fields, however, the ordering region is broadened and moves to significantly higher temperatures with increasing fields.

The magnetic properties of solid ^{3}He have been discussed for many years in terms of the Heisenberg nearest neighbor model (HNN). Most of the early high-temperature or low-magnetic field experiments were found to be compatible with this model and indicated an antiferromagnetic ordering of the nuclear spins at 2.0 mK at the melting pressure.[1] More recent results, however, seem to indicate that the simple HNN model is inadequate.[2,3,4] The effect of a high magnetic field on the thermodynamic pressure of solid ^{3}He was found to be about a factor of two lower than predicted by HNN.[2] In more recent observations a magnetic transition in solid ^{3}He on the melting curve was found to occur at 1.10 mK, almost a factor of two lower than expected.[4] Although attempts have been made to reconcile these

---

*Work supported in part by the National Science Foundation and the Research Corporation.

†Present address, Department of Physics, Colorado State University, Fort Collins, Colorado.

‡John Simon Guggenheim Memorial Fellow on leave from Cornell University, September 1974 - June 1975.

results with the Heisenberg model by including higher order exchange terms,[5] there presently exists no theory which successfully explains all the experimentally observed properties of solid ^{3}He.

In order to provide more information on the nature of the magnetic spin ordering in solid ^{3}He, we have studied the ordering in applied magnetic fields up to 1.2 T. We find that for $B \lesssim 0.42$ T the transition is similar to that reported by Halperin et al. in zero field.[4] The transition is sharp (i.e. covers a very narrow temperature interval) and is depressed in temperature by the field as is normally expected for an antiferromagnetic or spin-flop transition. At 0.42 T and above the ordering region broadens and moves to higher temperatures with increasing fields. For 1.2 T the ordering temperature has increased to 1.60 mK.

The experiment was performed in a Pomeranchuk adiabatic compression cell with the magnetic field provided by a superconducting solenoid. Further details of the apparatus and techniques have been described elsewhere.[6] The method which we have used to study the solid ordering is essentially the same as that of Halperin et al.[4] With the temperature held constant by the appropriate compression rate, an amount of heat ΔQ is added to the liquid and solid ^{3}He mixture, causing a change in volume ΔV as liquid is converted to solid. The quantities ΔQ and ΔV are measured as a function of the melting pressure P. In order to determine the temperature, use is made of the Clausius-Clapeyron equation

$$dP/dT = \Delta S/\Delta V = \Delta Q/(T\Delta V)$$

to obtain

$$\int_{T_o}^{T} T^{-1} dT = \int_{P_o}^{P} (\Delta V/\Delta Q)\, dP$$

where T_o and P_o are taken at a point where the melting pressure has been reasonably well established.[7] With $\Delta V/\Delta Q$ measured as a function of P, evaluating the integral gives T(P). Next the Clausius-Clapeyron equation is used to give the solid entropy $S_s(T)$, using the known behavior of the liquid entropy and the molar volume difference.[7,9]

On Fig. 1 are plots of $\Delta Q/\Delta V$ (= T dP/dT) for four different magnetic fields. As expected, in higher fields T dP/dT falls gradually below that for lower fields reflecting the depression of the melting curve by the field. In low fields a pressure is reached where T dP/dT undergoes a rapid decrease with increasing pressure (this feature is identified with the onset of ordering in the solid). For $B \lesssim 0.42$ T it happens that the ordering occurs at essentially the same pressure in each field. Since the melting

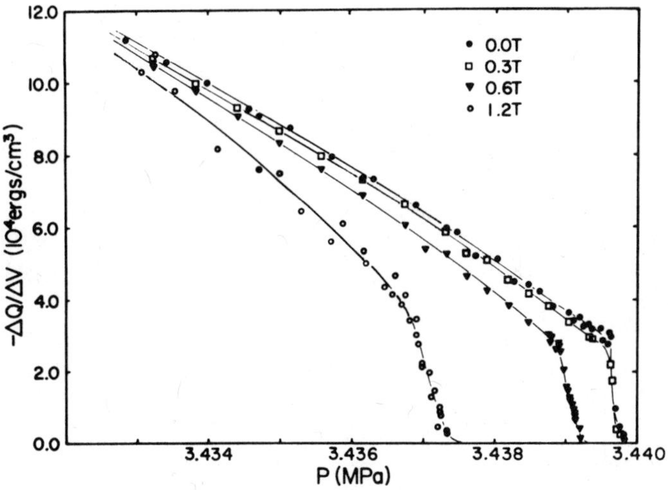

Figure 1. Raw data $\Delta Q/\Delta V = TdP/dT$ versus P for various magnetic fields [1 Pa (Pascal) = 10 μbar]. The plot covers the region below about 3.0 mK. The lines are "eyeball" fits to the points and were used to generate the required density of points for the numerical integration.

pressure is depressed in a field this indicates that the transition temperature is progressively lowered by increasing fields in this range. In higher fields the character of the ordering has changed dramatically (apparently discontinuously for $0.40 < B \leq 0.42$ T), with the ordering region becoming much broader and decreasing in pressure with increasing fields.

From the data of Fig. 1, T(P,B) and $S_s(T,B)$ were determined as discussed above (see Fig. 2). Our zero field data are consistent with that of Halperin, et al.[4,7] within our combined uncertainties. As a comparison of our overall temperature scales we find the A and B' transitions in liquid ^{3}He to be at $T_A = 2.68$ mK and $T_{B'} = 2.10$ mK. The accuracy of our temperatures is estimated to be about 8% with the precision somewhat better than this.

Well above the transition, there is a decrease in the entropy in a field relative to that for zero field because of spin alignment by the applied field. For low fields the entropy drops precipitiously at $T_c \simeq 1$ mK. Upon increasing the field, T_c decreases for $B \leq 0.4$ T. The behavior of $S_s(T,B)$ is completely different for $B \geq 0.42$ T. Now the reduction in entropy occurs much more gradually, and the ordering temperature, defined by the maximum in the specific heat,

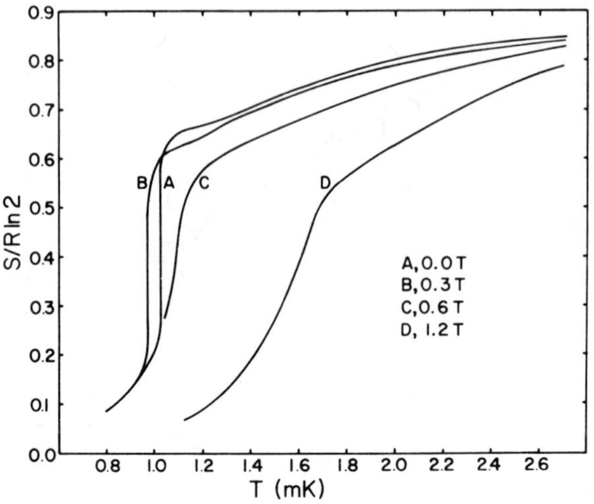

Figure 2. Entropy versus temperature for various magnetic fields, as determined from the lines in Fig. 1.

C = TdS/dT, increases with increasing fields. These conclusions are each related to the features of the TdP/dT data of Fig. 1 and are supported qualitatively by these raw data.

The ordering temperature versus field is plotted in Fig. (3a). Although a sufficient number of points have not yet been accumulated to show the detailed shape of the curves (each point requires about 10 days running time), it is clear that the low- and high-field points fall on two different branches. Another way of showing qualitatively that the character of the transition has changed above 0.42 T is provided by Fig. (3b). Here we have plotted $P_{max}(0) - P_{max}(B)$ versus B^2, where P_{max} is the maximum melting pressure reached in the compression. Compressions have been performed in several fields between 0.3 and 0.6 T (for which TdP/dT data are incomplete) to determine where the change from low-field to high-field behavior occurs. For $B \leq 0.40$ T, $P_{max}(0) - P_{max}(B) \simeq 0$, but at 0.42 T there is a rapid rise to the previously observed B^2 dependence of this quantity in higher fields.[10] A more dramatic change occurs in the "width" of the transition. This is a qualitative feature which we take here to be the difference in P_{max} for two particular compression rates differing by about a factor of five. The inset of Fig. (3b) shows the width versus B^2. There appears to be a discontinuous or very rapid change in width for $0.4 < B \leq 0.42$ T.

The significance of the two branches of the curve in Fig. (3a) is not yet fully understood. Solid ^{3}He is expected to have a very

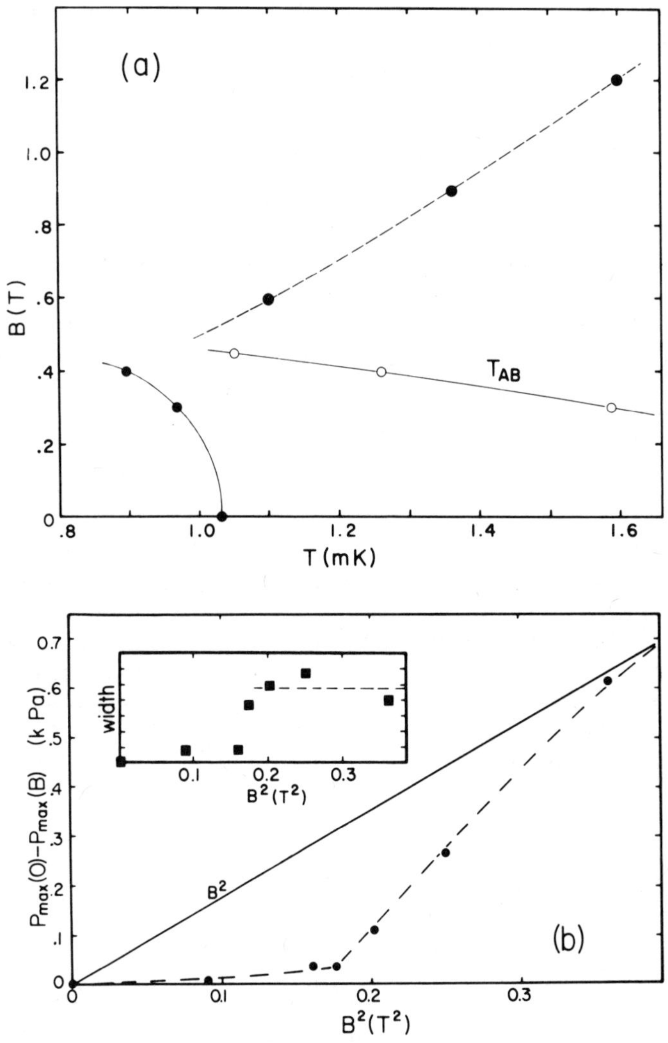

Figure 3. (a) Magnetic field versus ordering temperature, closed circles. The ordering temperature is defined by the maximum in the specific heat C = TdS/dT. The detailed shape of the curves is not known; the lines are merely suggestive. The open circles show our location of the BA transition between the superfluid phases of liquid ^{3}He. (b) Field dependence of P_{max}, the maximum pressure reached in compression. The B^2 line shows the field dependence found at higher fields. Inset: "width" of the transition (relative scale) versus B^2.

small anisotropy field arising from weak dipolar interactions. Thus, if the ordering in zero field were antiferromagnetic, only a small field (less than 0.1 mT) would be required to produce a spin-flop phase. It would be reasonable to assume, then, that the lower curve in Fig. (3a) is the paramagnetic-to-spin-flop boundary. The curvature of this line seems to indicate a critical field of less than 1.0 T--considerably less than the value 7.4 T predicted by the HNN model on the basis of earlier experiments.

The upper curve in Fig. (3a) may be associated with paramagnetic ordering of the ^{3}He spins by the magnetic field. The shape of the entropy versus temperature curves obtained for these fields shows some similarity to the corresponding curves for a non-interacting paramagnetic system in a field about 1.2 T greater than the actual applied field. This would suggest that the interaction of the spins produces an effective field parallel to the external field. The ordering region for a non-interacting paramagnet, however, is much broader than that which we have observed for solid ^{3}He.

Another curious feature seen in Fig. (3a) is that the curve T_{AB} corresponding to the transition between the two superfluid phases of liquid ^{3}He seems to intersect the solid transition line about 0.45 T, very near the field at which the solid changes behavior. It is unknown whether this is a coincidence or whether there exists a relationship between the AB transition in liquid ^{3}He and the magnetic ordering in solid ^{3}He. We believe the latter possibility is unlikely, although it should be remarked that the measured quantities relate to phenomena occurring at the liquid-solid interface and that the bulk solid is not in thermal equilibrium with the liquid.

References

(1) S. B. Trickey, W. P. Kirk, and E. D. Adams, Rev. Mod. Phys. 44, 668 (1972).

(2) W. P. Kirk and E. D. Adams, Phys. Rev. Lett. 27, 392 (1971).

(3) J. M. Dundon and J. M. Goodkind, Phys. Rev. Lett. 32, 1343 (1974)

(4) W. P. Halperin, C. N. Archie, F. B. Rasmussen, R. A. Buhrman, and R. C. Richardson, Phys. Rev. Lett. 32, 927 (1974).

(5) L. I. Zane, J. Low Temp. Phys. 9, 219 (1972); A. K. McMahan and R. A. Guyer, Phys. Rev. A7, 1104 (1973); L. Goldstein, Phys. Rev. A8, 2160 (1973); and L. I. Zane and J. R. Sites, J. Low Temp. Phys. 17, 159 (1974).

(6) R. B. Kummer, E. D. Adams, W. P. Kirk, A. S. Greenberg, R. M. Mueller, C. V. Britton, and D. M. Lee, Phys. Rev. Lett. 34, 517 (1975).

(7) W. P. Halperin, Ph.D. thesis, Cornell University (unpublished). We have used T_o = 16.88 mK and P_o = 3.37830 MPa. Our pressure calibration was adjusted so that the A transition in the liquid for B = 0 occurs at P_A = 3.43420 MPa.

(8) W. R. Abel, A. C. Anderson, W. C. Black, and J. C. Wheatley, Phys. Rev. 147, 111 (1966).

(9) R. A. Scribner, M. F. Panczyk, and E. D. Adams, J. Low Temp. Phys. 1, 313 (1969); and E. R. Grilly, ibid. 4, 615 (1971).

(10) D. D. Osheroff, R. C. Richardson, and D. M. Lee, Phys. Rev. Lett. 28, 885 (1972).

THEORETICAL SEARCH FOR COLLISIONLESS ORBIT WAVES

W. M. Saslow

Department of Physics, Texas A & M University

College Station, Texas 77843

Abstract: Combescot's form of the kinetic equation is studied in the collisionless regime in a search for orbit waves in He^3- A. No solutions are found for real values of the velocity. It is pointed out that spontaneously broken symmetry does not imply an associated mode in the collisionless regime.

One of the most interesting aspects of the identification of He^3- A as an $\ell = 1$ pairing system is that one expects a new mode associated with motion of the axis of orbital quantization.[1] Work has already been done employing Ginzburg-Landau theory (Wölfle)[2] and hydrodynamic theory (Graham)[3] to this new mode, or orbit wave. The former theory has the advantage that, near T_c, it explicitly gives the dispersion relation in the absence of damping. The latter theory has the advantage that it gives the structure of the dispersion relation, including damping, but the disadvantage that none of the constants appearing in the theory are known. Ideally, one should determine the properties of this mode by studying the transport equation.

To do this, one would have to determine the "quasi-conserved" quantities associated with the 4 x 4 quasiparticle density matrix $\delta\nu$,[4] in order to develop the proper conservation laws for the quantity associated with the motion of the gap axis. This can be expected to be a rather complex procedure. As a start, therefore, I have considered the related (but simpler) problem of high-frequency orbit waves. I find that there are good reasons to believe that the equation describing these modes has no propagating solutions.

One starts from Combescot's equations[5] for the change in the anomalous part of the energy matrix $E_{-,k}$, due to a disturbance of frequency ω and wave-vector $\vec{q}$:

$$\delta E_{-,k} = \delta\Delta_k = \sum_{k'} V_{kk'} \delta n_{-,k'} .$$

Here k is the quasi-particle wave-vector, $V_{kk'} \approx 3V_1 \vec{k}\cdot\vec{k}'$, $\delta\Delta_k$ is the change in the gap function Δ_k, and $\delta n_{-,k'}$ is the anomalous part of the "non-diagonalized" quasi-particle density matrix.

Through a series of transformations which Combescot has already discussed, one:

1) goes from $\delta n_{-,k'}$ to the "diagonalized" quasi-particle density matrix $\delta\nu_{k'}$;
2) employs the kinetic equation which relates $\delta\nu_{k'}$ to the "diagonalized" energy matrix $\delta E_{k'}$; and
3) goes from $\delta E_{k'}$ to $\delta\epsilon_{k'}$.

One finds the following equation for $\delta\Delta_k$. Note that the non-anomalous parts do not couple to $\delta\Delta_k$, but I have not studied whether or not a similar equation for the non-anomalous parts will connect to $\delta\Delta_k$. This would not matter in determining the dispersion relation for this mode, but it would be relevant to the problem of determining how the mode couples to observable quantities.

For a general $\delta\Delta_k$ we find (in the absence of Fermi liquid corrections) that

$$\delta\Delta_k = \sum_{k'} V_{kk'} \left(\frac{\phi}{E}\right)_{k'} \left[-\frac{(\Delta^+\delta\Delta)}{2E^2} + \frac{2\xi^2+|\Delta|^2}{2E^2}\,\delta\Delta \right]_{k'}$$

$$+ \sum_{k'} V_{kk'} \left[\frac{\Delta\delta\nu_e - \delta\nu_t\Delta}{2E} \right]_{k'} .$$

In the above, $\phi_k = -\frac{1}{2}\tanh\frac{1}{2}\beta E_k$, where $\beta = (k_B T)^{-1}$ and E_k is the quasi-particle energy, ξ_k is the kinetic energy measured from the Fermi energy, and $\delta\nu_{e,t}$ are defined in Ref. 5.

Since the equilibrium gap is given, for the axial phase, by

$$\Delta_k = \Delta_o(\vec{k}_x + i\vec{k}_y),$$

where Δ_o is the maximum value of the equilibrium gap and we have not included the spin indices (it is diagonal in them), a rotation of the gap axis $\vec{\ell}_z$ gives[2]

$$\delta\Delta_k = \Delta_o \vec{k}_z(-i\delta\vec{\ell}_x + \delta\vec{\ell}_y) .$$

Inserting this into the "gap" equation, eliminating $\delta\nu$ by use of the kinetic equation, and employing the p-wave symmetry of $V_{kk'}$ one finds that

$$\delta\Delta_k = \delta\Delta_k \{I_1 + \frac{3}{2} I_2(s)\}V_1$$

where $s = (\omega/v_F q)$ and

$$I_1 = 3 \sum_k \left(\frac{\phi}{2E^3}\right)_k \vec{k}_z^2 \, [E^2+|\Delta|^2]_k ,$$

$$I_2(s) = - \sum_k \phi_k' \left(\frac{|\Delta|^2}{E}\right)_k \vec{k}_z \left|\frac{(\vec{q}\cdot\vec{k}\xi/E)^2}{s^2-(\vec{q}\cdot\vec{k}\xi/E)^2}\right|$$

Here v_F is the Fermi velocity, and $\phi_k' = \delta\phi_k/\delta E_k$. Using the equilibrium gap equation

$$1 = \frac{3V_1}{2\Delta_o^2} \sum_k \left(\frac{\phi}{E}\right)_k |\Delta|_k^2$$

this can be rewritten as

$$0 = \delta\Delta_k[I_o - I_2(s)]$$

where

$$I_o = \sum_k \left(\frac{\phi}{E}\right)_k \left[(1-3\vec{k}_z^2) + \vec{k}_z^2 \frac{\xi_k^2}{E_k^2}\right] .$$

Hence, $I_o = I_2(s)$ for some value of s, or $\delta\Delta_k = 0$. One can now search for solutions to this equation.

Note that ϕ' and $\frac{\phi}{E}$ are both negative and that $\phi' < \frac{\phi}{E}$, since

$$\frac{\phi}{E} - \phi' = -E \left(\frac{\phi}{E}\right)' = \frac{1}{2} \beta E \left(\frac{\tanh \frac{1}{2} \beta E}{E}\right)' < 0 .$$

Therefore $I_o < 0$, since the bracketed term is generally positive. To get $I_2(s) < 0$ also, we must have $s^2 < 1$. Further, $I_2(0)$ is the most negative value of I_2 for real s. We therefore compare I_o and $I_2(0)$. We have

$$I_o \approx \sum_k \left(\frac{\phi}{E}\right)_k \vec{k}_z^2 \frac{\xi_k^2}{E_k^2} ,$$

since $\langle 1 - 3\vec{k}_z^2\rangle \approx 0$ for most of the range of integration, where $|\Delta|_k^2 \ll \xi_k^2$ (this is exactly true near T_c), and

$$I_2(0) = \sum_k \phi_k' \vec{k}_z^2 \left(\frac{|\Delta|^2}{E^2}\right)_k .$$

Since $\xi_k^2 \gg |\Delta|_k$ over most of the range of integration, we have $|(\phi/E)\xi^2| > |\phi'|\Delta|^2|$, and $|I_0| > |I_2(0)|$. Therefore, we expect no solution. Certainly there is none for T = 0 and $T \to T_c$. I have performed a few numerical searches ($\beta\Delta_o$ = 0.2, 1.0, 2.0), finding no solutions.

From this it appears that, unless Fermi liquid corrections are quite significant, the collisionless orbit wave will not propagate. I have not searched for solutions with arbitrary complex values of s, but it can readily be verified that there are no solutions for imaginary s.

In conclusion, it should be observed that the absence of a collisionless orbit wave does not contradict the principle that, associated with each continuously broken symmetry, (here, the direction of $\vec{\ell}$) there is a collective mode with the property that $\omega \to 0$ as $q \to 0$.[6] For, in the case of collisionless waves, one cannot take the $\omega \to 0$ limit because the collisionless equations become invalid in that limit, where collisions must dominate.

References

(1) P. W. Anderson, Phys. Rev. Lett. 30, 368 (1973).

(2) P. W. Wölfle, Phys. Lett. 47A, 224 (1974).

(3) R. Graham, Phys. Rev. Lett. 33, 1431 (1974).

(4) O. Betbeder-Matibet and P. Nozieres, Ann. Phys. (New York) 51, 392 (1969).

(5) R. Combescot, Phys. Rev. A 10, 1700 (1974).

(6). P. W. Anderson, Concepts in Solids (Benjamin, N. Y., 1963).

SPIN DYNAMICS IN SUPERFLUID ^3He

R. Combescot

Laboratory of Atomic and Solid State Physics,
Cornell University
Ithaca, New York 14853

Abstract: An approach to spin dynamics is presented, which is based on a generalization of the Landau-Silin kinetic equation to the superfluid state. This method allows us to treat the nonlinear regime as well as the linear one.

We would like first to outline the principle of the formalism that we are using to study spin dynamics in superfluid ^3He for frequencies small compared to the gap: $\omega << \Delta$. We will then give some results on the longitudinal NMR in the A phase and on the generalization of this formalism to the nonlinear regime.

The basic equation in this formalism is a kinetic equation, or a Boltzmann equation, for the quasi-particle distribution derived within the BCS theory. In the regime $\omega << \Delta$, quasi-particles are not created or destroyed and the kinetic equation describes the evolution of the quasi-distribution from its equilibrium value given by the Fermi distribution. In the normal state, this equation is merely the Landau kinetic equation (or the Landau-Silin equation if we include the magnetic field). In the superfluid state, one can similarly write a kinetic equation for the quasi-particle distribution:

$$\frac{\partial \nu_k}{\partial t} + \frac{\partial \nu_k}{\partial \vec{r}} \frac{\partial E_k}{\partial \vec{k}} - \frac{\partial \nu_k}{\partial \vec{k}} \frac{\partial E_k}{\partial \vec{r}} = I(\nu_k) \ , \qquad (1)$$

where ν_k is the quasi-particle distribution, E_k is the quasi-particle energy, and $I(\nu_k)$ the collision term. But it happens that this is possible only for a special choice of the gauge; whereas, in the normal state, the gauge is naturally completely free. By gauge transformation, we mean a transformation corresponding to the multiplication of the wave functions by exp $[i\sigma_i\theta_i(r,t)]$, that is, a gauge transformation in the usual meaning together with a time- and space-dependent spin rotation. This choice of the gauge is naturally linked to the fact that the superfluid state is breaking the gauge invariance. This problem is discussed in detail by Betbeder-Matibet and Nozieres.[1]

In the low frequency regime, the degrees of freedom of the order parameter are strongly reduced since any change of the structure of the order parameter is expected to correspond to frequencies $\sim \Delta$. If we are interested in sound propagation or spin dynamics, we are left with phase changes and rotation in spin space (since the order parameter is a vector in spin space). The gauge which must be chosen for the kinetic equation (1) is such that after the transformation, these phase changes and spin rotation of the order parameter are canceled. In other words, all the space-time dependence of the order parameter is included in the gauge transformation. I would like to stress the physical meaning of $\theta_i(r,t)$, which is made of: 1) $\theta_o(\vec{r},t)$ corresponding to the unit Pauli matrix and which is the usual phase, and 2) $\vec{\theta}(\vec{r},t)$ corresponding to σ_x, σ_y, σ_z, which is some generalized phase. If one knows $\vec{\theta}(\vec{r},t)$, one knows where the order parameter is pointing in spin space.

Another physical interpretation is related to the two fluid model. One of the main advantages of this formulation is precisely that it can be interpreted within a two fluid model which is a very appealing one. The quasi-particles correspond to the normal fluid, and the superfluid is described by θ_o and $\vec{\theta}$. The kinetic equation describes how the normal fluid relaxes toward the equilibrium defined by the superfluid parameters. From Josephson's equation, we know that the chemical potential of the superfluid is related to the phase by:

$$\delta\mu = -\frac{\partial\theta_o}{\partial t} \quad . \tag{2}$$

In the same way, we have a difference in the chemical potential between up and down spins:

$$\overrightarrow{\delta\mu} = -\frac{\partial\vec{\theta}}{\partial t} \quad , \tag{3}$$

if the spin quantization axis is taken along $\overrightarrow{\delta\mu}$. We have assumed for the moment that $\vec{\theta}$ is small in writing Eq. (3). Alternatively, one can say that the superfluid is imposing an effective field $\overrightarrow{\delta\mu}$ to the normal fluid. We know also that the superfluid velocity $\vec{v}_s$ is related to the gradient of the phase θ_o through

$$\vec{v}_s = \frac{1}{m}\,\overrightarrow{\text{grad}}\;\theta_o \quad . \tag{4}$$

Here in the same way, the superfluid will have, in addition, a spin superfluid velocity

$$\vec{\vec{v}}_s = \frac{1}{m_s} \overrightarrow{\text{grad}}\ \vec{\theta} \tag{5}$$

where m_s is, in general, different from m. $\vec{\vec{v}}_s$ is a vector in both orbital space and spin space, and tells simply that there is a difference between spin up and spin down superfluid velocity.

We can linearize the kinetic equation (1) since the departure of the distribution function from equilibrium is always small. Therefore:

$$\frac{\partial E_k}{\partial \vec{k}} = \frac{\partial}{\partial \vec{k}} (\xi_k^2 + |\Delta_k|^2)^{1/2} \text{and } \frac{\partial \nu_k}{\partial \vec{k}} = \frac{\partial f}{\partial \vec{k}} = f' \frac{\partial E_k}{\partial \vec{k}} \tag{6}$$

where f' is the derivative of the Fermi distribution f. Eq. (1) becomes:

$$\frac{\partial (\partial \nu_k)}{\partial t} + \frac{\partial E_k}{\partial \vec{k}} \frac{\partial}{\partial \vec{r}} (\delta \nu_k - f' \delta E_k) = I(\nu_k) \ . \tag{7}$$

In this equation, $\delta\nu_k$ is the departure of the quasi-particle distribution from equilibrium and δE_k is the local change in quasi-particle energy due to the superfluid and to Fermi liquid effects. Here, $\delta\nu_k$ and δE_k are both 2x2 matrices in spin space. The part proportional to the unit matrix corresponds to fluctuations in density and the rest corresponds to fluctuations in magnetization.

We still need an equation to determine θ_0 and $\vec{\theta}$ to be able to use the kinetic equation. This is done by writing particle conservation and spin conservation which are not automatically satisfied by the kinetic equation because of the quasi-classical expansion. This point is discussed in detail in Betbeder-Matibet and Nozieres.[1]

Particle conservation gives:

$$\frac{\partial}{\partial t} \delta\rho + \text{div}\ \vec{j} = 0 \tag{8}$$

where $\delta\rho$ is the change in density and $\vec{j}$ the particle current. To obtain $\vec{\theta}$, we must write spin conservation, which is

$$\frac{\partial}{\partial t} \overrightarrow{\delta\rho} + \text{div}\ \vec{\vec{j}} = D(\vec{\theta}) \tag{9}$$

where $\delta\vec{\rho}$ is the fluctuation in magnetization and $\vec{j}$ the spin current (vector both in orbital and spin space). We know from Leggett[2] that the dipole interaction breaks the spin conservation and we have in the right-hand side a term $D(\vec{\theta})$ coming from the dipole interaction which has been derived by Leggett. If we consider the linear regime ($\vec{\theta}$ small), this term is $\overleftrightarrow{\phi}^{(o)}\vec{\theta}$ where $\overleftrightarrow{\phi}^{(o)}$ has been introduced by Leggett. This conservation law is simply obtained by writing $\frac{\partial}{\partial t}\delta\vec{\rho} = [\delta\vec{\rho},H]$ and performing the spin rotation. It is not basically different from the first Leggett equation, except that space dependence is also introduced.

We now have everything to study spin dynamics; but if we do not want to be restricted to the collisionless or to the hydrodynamic regime, we need to specify the collision term. What we have done is to approximate this collision term by a relaxation time approximation. Since local equilibrium corresponds to $\delta\nu_k = f'\delta E_k$ and collisions are producing local equilibrium, we choose (for the spin part):

$$I(\delta\nu_k) = -\frac{(\delta\nu_k - f'\delta E_k)}{\tau} \quad . \tag{10}$$

This approximation satisfies automatically the conservation law, Eq. (9), which is explicitly required in this formalism, and reduces, in the normal state, to a usual (spin conserving) relaxation time approximation, although not in a gauge invariant way. However, it has been pointed out recently by Leggett[3] and Ambegaokar[6] that, for the study of the NMR, the relaxation time $\tau(T)$ cannot be taken, near T_c, as the spin diffusion relaxation time in the normal state, as assumed before; $\tau(T)$ must actually diverge near T_c. This can be understood by looking at the relaxation processes in the normal state. For the NMR, the question is how a quasi-particle distribution, uniform over the Fermi surface, is reacting to a change in chemical potential between up spin and down spin. The answer is that, because the distribution is uniform and the collisions between quasi-particles conserve the spin, the collision term is identically zero, which means that the corresponding relaxation time must be infinite in the normal state. For spin diffusion calculation,[5] on the other hand, the relaxation time is finite in the normal state and its value near T_c in the superfluid state should not be different from its value in the normal state.

For the longitudinal NMR linewidth in the phase, one obtains:[4]

$$\Delta\omega = \frac{\omega_o^2}{1 + F_o^a}\,\tau(T)f(T) \qquad f(T) = \int \frac{d\Omega}{4r}\,d\xi(-f')\left(\frac{\xi}{E}\right)^2 \tag{11}$$

where ω_o is the longitudinal resonance frequency, F_o^a the s-wave antisymmetric Fermi liquid parameter and $\tau(T)$ is the relaxation time discussed above. An interesting problem is the limit of this linewidth when the temperature is going to zero. To answer that question, one must go, in general, beyond the hydrodynamic assumption $\omega_o \tau(T) \ll 1$ which has been used to derive Eq. (11). It turns out that the linewidth is going to zero at zero temperature whatever the behavior of $\tau(T)$ at low temperature. One can understand this result in the following way: either $\omega_o \tau(T) \ll 1$ is still true at low temperature and the linewidth goes to zero according to Eq. (11), or $\omega_o \tau(T) \gg 1$ at low temperature and one goes to the collisionless régime where the linewidth is proportional to $f(T)/\omega_o \tau(T)$ and goes again to zero. This result for the NMR linewidth should hold also for the isotropic state and for the transverse linewidth.

We can apply exactly the same formalism to study the nonlinear regime for the longitudinal resonance, for example, the ringing experiments. The only modification is that the dipole term in Eq. (9) is no longer linear. In this regime, we derive the following differential equation for the phase difference θ between up spins and down spins:

$$\tau(1-f(T))\,\dddot{\theta} + \ddot{\theta} + \dot{\theta}\,\tau\omega_o^2\cos(2\theta)\left(1 - \frac{F_o^a f(T)}{1 + F_o^a}\right) + \frac{\omega_o^2}{2}\sin 2\theta = 0 \qquad (12)$$

This equation has also been derived by V. Ambegaokar.[6]

More generally, the preceding formalism can be generalized to study space- and time-dependent spin dynamics in the nonlinear regime. This extension is rather straightforward because the main point of this theory is the low frequency, long wavelength expansion, which is essentially not modified in the nonlinear regime. One main modification is that $\vec{\delta\mu}$ and $\vec{\vec{v}}_s$ must now be defined as:

$$\vec{\delta\mu} = \frac{i}{2}\,\mathrm{Tr}\,\{\vec{\sigma}\, e^{-i\vec{\sigma}\cdot\vec{\theta}}\,\frac{\partial}{\partial t}\, e^{i\vec{\sigma}\cdot\vec{\theta}}\}$$

$$\vec{\vec{v}}_s = -\frac{i}{2m^*}\,\mathrm{Tr}\,\{\vec{\sigma}\, e^{-i\vec{\sigma}\cdot\vec{\theta}}\,\frac{\partial}{\partial\vec{r}}\, e^{i\vec{\sigma}\cdot\vec{\theta}}\}. \qquad (13)$$

Using this theory, one can rederive in the hydrodynamic regime, the second Leggett equation:

$$\frac{\partial}{\partial t}\vec{d}_k = \vec{d}_k \times (\vec{H} - \bar{\chi}^{-1}\,\vec{\delta\rho}) \qquad (14)$$

where $\overline{\overline{\chi}}$ is the static susceptibility tensor. One can also derive a similar equation for the space variation of the order parameter $\vec{d}_k$

$$\frac{\partial}{\partial x_i} \vec{d}_k = \vec{d}_k \times [(\overline{\overline{\rho}}_s)^{-1}_{ij} \vec{j}_j] \tag{15}$$

where $\overline{\overline{\rho}}_s$ is the spin superfluid density tensor and $\vec{j}_j$ the spin current.

In conclusion, our method provides a general approach to spin dynamics which allows us to consider any frequency regime (provided that the frequency is small compared to the gap). Up to now, this method has been used only to treat the linear regime; but it can be extended rather easily to treat the nonlinear regime, where the direction of the order parameter has large deviations from the equilibrium position.

References

(1) O. Betbeder-Matibet and P. Nozieres, Ann. Phys. 51, 392 (1969).

(2) A. J. Leggett, Phys. Rev. Lett. 31, 352 (1973).

(3) A. J. Leggett, to be published.

(4) R. Combescot and H. Ebisawa, Phys. Rev. Lett. 33, 810 (1974).

(5) R. Combescot, Phys. Rev. Lett. 34, 8 (1975).

(6) V. Ambegaokar, to be published.

SPIN WAVES AND MAGNETIZATION OSCILLATION IN SUPERFLUID ^{3}He

Kazumi Maki

Department of Physics, University of Southern California

Los Angeles, California 90007

I. INTRODUCTION

Since a pioneering NMR experiment by Osheroff et al,[1] it has been gradually realized that the superfluid phases of liquid ^{3}He possess extraordinary magnetic properties. All of the extraordinary magnetic properties reflect the fact that the condensates in both the A and B phases of superfluid ^{3}He comprise the triplet pairs as first recognized by Leggett.[2,3] The ground state of superfluid ^{3}He is highly degenerate as to the spin configuration as well as the orbital configuration of the triplet pairs. This implies that the superfluid ^{3}He possesses a class of gapless collective modes associated with the rotation of the spin configuration (i.e., Goldstone boson). As pointed out by Leggett[2,3] the dipole interaction energy between nuclear spins of ^{3}He atoms, though extremely small, will lift partially the high degeneracy of the ground state. The dipole interaction introduces correlations between the orbital and the spin configuration of the condensate, giving rise to an additional shift in the NMR frequencies for example.

We will review here some of the work done with Tsuneto and with Hu as well as my own work on motion of spin configurations in superfluid ^{3}He.[4-7] In the following we will consider the A phase and the B phase separately, since we will employ different methods in deducing the basic dynamical equations governing the motion of spin configurations in these two phases; in the A phase we exploit an analogy between the longitudinal resonance and the Josephson effect,[4,5] while in the B phase we consider the general gauge transformation associated with the rotation of the spin configuration.[6] Then, limiting ourselves to small oscillations around the equilibrium configuration, we predict the existence of spin waves in both phases of superfluid ^{3}He, which are considered as the Goldstone boson associated with the rotation of the spin configuration; the spin wave dispersion is determined, which has, in general, a nonvanishing energy gap associated with the dipole interaction energy except in special configurations. We will also discuss the ringing of magnetization after a sudden application or removal of magnetic field, which is compared with recent experiments by Webb et al.[8,9]

II. SPIN DYNAMICS IN THE A PHASE

Since the condensate of the A phase is described in terms of the axial state,[10] we can imagine that the A phase consists of two superfluids described by $\Delta_\uparrow$ and $\Delta_\downarrow$ in the presence of a magnetic field H along the Z axis. Here $\Delta_\uparrow$ and $\Delta_\downarrow$ are order parameters associated with the up spin atom pairs and the down spin atom pairs. In the absence of the dipole interaction these two superfluids are completely independent of each other. The dipole interaction introduces a weak coupling between $\Delta_\uparrow$ and $\Delta_\downarrow$; the energy of the total system now depends on the relative phase $\phi = \phi_\uparrow - \phi_\downarrow$ where $\phi_\uparrow$ and $\phi_\downarrow$ are the phases of $\Delta_\uparrow$ and $\Delta_\downarrow$ respectively. As in the case of the Josephson junction we can introduce the conjugate variable n by:

$$[n, \phi] = -i \tag{1}$$

where n is expressed in terms of the number of the up spin atoms and the down spin atoms as

$$n = 1/4\ (n_\uparrow - n_\downarrow) = 1/2 S_z \tag{2}$$

the half of the z component of the spin density (therefore proportional to the z component of magnetization M_z).

The total Hamiltonian of the system is given[4,5] as

$$H = \varepsilon_\uparrow + \varepsilon_\downarrow + \varepsilon_I + \varepsilon_d \tag{3}$$

where $\varepsilon_\uparrow$ and $\varepsilon_\downarrow$ describe free motion of the up spin quasi-particles and the down spin quasi-particles respectively, ε_I is the spin exchange interaction,

$$\varepsilon_I = I \int n_\uparrow(\vec{r}) n_\downarrow(\vec{r}) d^3r \ , \tag{4}$$

ε_d is the dipole interaction energy

$$\varepsilon_d/V = H_d = -\frac{\pi \gamma_0^2}{20 g^2} \Delta^2 \ (1 + 3\cos\phi) \ , \tag{5}$$

γ_0 is the gyromagnetic ratio of ^{3}He nucleus, and g is the pairing interaction constant. Here we assume that the condensate is described by the P-wave triplet pairs. From the total spin conservation we have then[5]

$$\dot{S}_z + \vec{\nabla}\vec{j}_{sz} = 2\left(\frac{\partial H}{\partial \phi}\right) = \frac{3\pi\gamma_o^2}{10g^2}\,\Delta^2 \sin\phi \tag{6}$$

where $\vec{j}_{sz}$ is the spin current and a dot means the time derivative. In the absence of the dipole interaction or in the normal state, the rhs of Eq. (6) vanishes identically as it should. In the hydrodynamic limit, the spin current is expressed in terms of the gradient of ϕ as,[5]

$$\vec{j}_{sz} = \frac{N}{8m}\,(\overleftrightarrow{\rho_s}/\rho)\,\vec{\nabla}\phi \tag{7}$$

where $(\overleftrightarrow{\rho_s}/\rho)$ is identical to the reduced superfluid density tensor (associated with the mass flow).[5] In order to complete Eq. (6), we need a relation between $\dot{\phi}$ and n (or S_z), which is provided by the Josephson relation

$$\dot{\phi} = -\frac{\delta H}{\delta n} = -2(\mu_\uparrow - \mu_\downarrow) + 4IS_z \tag{8}$$

where $\mu_\uparrow$ and $\mu_\downarrow$ are the chemical potential of the up spin atom and the down spin atom respectively. Finally, eliminating $\vec{S}_z$ and $\vec{J}s_z$ from Eq. (6) , we obtain

$$\ddot{\phi} - 1/3(1-\bar{I})v_F^2[(\vec{\nabla}\overleftrightarrow{\rho_s}\vec{\nabla})\phi]\rho^{-1} = -\Omega_A^2 \sin\phi \tag{9}$$

where $\Omega_A = [(1-\bar{I})6\pi\gamma_o^2/(5g^2N(0))]^{1/2}\,\Delta(T)$ the longitudinal resonance frequency of the A phase,[3] v_F is the Fermi velocity, and $\bar{I} = IN(0)$. Equation (9) is of fundamental importance in describing the motion of the longitudinal spins in the A phase.

If we consider a small oscillation in ϕ around the equilibrium configuration (ϕ=0), the oscillation propagates with the dispersion;

$$\omega^2 = \Omega_A^2 + 1/3(1-\bar{I})v_F^2(\overleftrightarrow{q\rho_s q})/\rho \tag{10}$$

At T=0K, the above dispersion agrees with a previous calculation in the collisionless limit by Maki and Ebisawa,[12] although at T ≠ 0K the present result differs significantly from the one in the collisionless limit. Recently, starting from the kinetic equation for quasi-particles in superfluid ^{3}He, Combescot[13] has determined the spin wave dispersion both in the hydrodynamic and the collisionless limit. His result in the hydrodynamic regime agrees with Eq. (10).

So far we have considered the longitudinal oscillation only. In the case of the transverse oscillation, the analogy to the

Josephson effect is useless, since in this case the spin oscillation involves an additional order parameter $\Delta_0 \propto \langle\psi_\uparrow\psi_\downarrow\rangle$. Therefore we will not go into details but point out that the fundamental equation is given by[5]

$$\ddot{\psi} - 1/3(1-\bar{I})v_F{}^2(\vec{\nabla}\overleftrightarrow{\rho_s}\vec{\nabla})\psi/\rho = -\omega_L{}^2\psi - \Omega_A{}^2 \sin\psi \tag{11}$$

where $\omega_L = \gamma_0 H$ the Larmor energy and ψ is twice the angle between $\vec{d}$ and $\vec{l}$, where $\vec{d}$ is the spin direction of the condensate and $\vec{l}$ is the direction of the angular momentum of the condensate. In particular Eq. (11) predicts a spin wave with the dispersion

$$\omega^2 = \omega_L{}^2 + \Omega_A{}^2 + 1/3(1-\bar{I})v_F{}^2\,(\vec{q}\overleftrightarrow{\rho_s}\vec{q})\rho^{-1} \tag{12}$$

So far we have neglected the dissipation term completely. More generally Eq. (9) will be formally generalized as

$$\ddot{\phi} - 1/3(1-\bar{I})v_F{}^2\,\vec{\nabla}\overleftrightarrow{\rho_s}\vec{\nabla}\phi/\rho - \vec{\nabla}\overleftrightarrow{D}\vec{\nabla}\phi - T_1{}^{-1}\phi = \Omega_A{}^2 \sin\phi \tag{13}$$

where $\overleftrightarrow{D}$ is the spin diffusion constant tensor and T_1 is the intrinsic-spin lifetime due to the quasi-particle scattering. Recent estimates[14-16] of T_1 and $\overleftrightarrow{D}$ appear to indicate that T_1 is negligible in most situations, while $\overleftrightarrow{D}$ may be rather important in the vicinity of the transition temperature.

III. RINGING OF MAGNETIZATION IN THE A PHASE

We will now consider a homogeneous situation. In this case Eq. (9) reduces to

$$\ddot{\phi} = -\Omega_A{}^2 \sin\phi \tag{14}$$

which is equivalent to that for a classical pendulum. When ϕ is small, Eq. (14) implies that ϕ oscillates with frequency Ω_A. We will study here the transient behavior of ϕ and $M_z(\equiv\gamma_0\chi S_z)$ after a sudden application of a magnetic field ΔH; we will solve Eq. (14) with the initial conditions

$$\phi = 0,\ \phi = 2(\Delta\omega_L) \text{ at } t = 0 \tag{15}$$

where $\Delta\omega_L = \gamma_0(\Delta H)$. Eq. (14) is integrated to give

$$\dot{\phi} = [(\Delta\omega_L)^2 - \Omega_A{}^2 \sin^2(\phi/2)]^{1/2} \tag{16}$$

and

$$S_z = 1/2\, N(0)(1-\bar{I})^{-1}[(\Delta\omega_L)^2 - \Omega_A{}^2 \sin^2(\phi/2)]^{1/2} \tag{17}$$

Furthermore Eq. (16) tells us that the longitudinal magnetization oscillates with a frequency ω_r;

$$\omega_r = \pi\Omega_A/[2K(\Delta\omega_L/\Omega_A)] \quad \text{for } \Delta\omega_L < \Omega_A$$

$$= \pi\Delta\omega_L/[K(\Omega_A/\Delta\omega_L)] \quad \text{for } \Delta\omega_L > \Omega_A \tag{18}$$

where K(z) is the complete elliptic integral. The ΔH dependence of ω_r is shown in Fig. 1. For small $\Delta\omega_L$, ω_r is Ω_A. As $\Delta\omega_L$ increases ω_r decreases and vanishes for $\Delta\omega_L = \Omega_A$. For $\Delta\omega_L > \Omega_A$, ω_r jumps up and approaches rapidly $\omega_r = 2(\Delta\omega_L)$ as $\Delta\omega_L$ increases. This behavior is easily interpreted in analogy with the motion of the pendulum where $\Delta\omega_L$ is the impulse imparted to the pendulum at t=0.

The remarkable nonlinear behavior predicted above has been observed in recent ringing experiment by Webb <u>et al</u>.;[8,9] they measured the ringing frequency of the magnetization after a sudden removal of the magnetic field ΔH and the observed frequency is at least in qualitative agreement with the present result.[9]

IV. GAUGE INVARIANCE AND SPIN WAVES IN THE B PHASE

We assume that the condensate of the B phase is described in

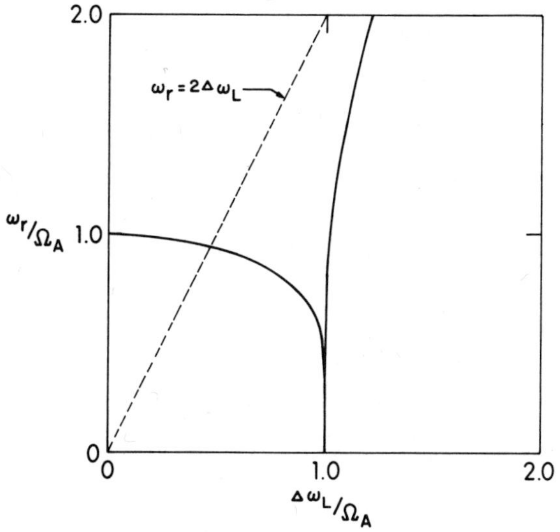

Figure 1. The ringing frequency ω_r in the nonlinear regime in the A phase plotted as a function of $\Delta\omega_L/\Omega_A$.

terms of a spherical triplet state of Balian and Werthamer,[17] since this choice appears most consistent with the known phase diagram of the superfluid ^{3}He.[18,19] As already noted by Leggett,[3] in the absence of the dipole energy, the ground state energy of the BW state is invariant against the following non-Abelian gauge transformation of the triplet order parameter $\vec{\Delta}(\Omega)$;

$$\vec{\Delta}(\Omega) \to e^{i\phi} R_\sigma(\alpha,\beta,\gamma) R(\acute{\alpha},\acute{\beta},\acute{\gamma}) \vec{\Delta}(\Omega) \tag{19}$$

where ϕ is the change in the overall phase factor, R_σ the rotation of the spin space, and R is the rotation of the coordinate space. Here $(\alpha\beta\gamma)$ and $(\acute{\alpha}\acute{\beta}\acute{\gamma})$ are Eulerian angles describing the rotation of the spin space and the coordinate space respectively. Therefore the BW state admits three different classes of Goldstone bosons:[7] zero sound,[20] spin wave,[7,13] and orbital wave. In the following we treat α, β, and γ, the local Eulerian angles describing the rotation of the spin space, as dynamical variables. In fact the spin polarization as well as spin current are expressed in terms of the time and the space derivative of the Eulerian angle;[6]

$$\begin{aligned} \vec{M} &= \gamma_0 \chi_B \vec{\omega} \\ \vec{J}_{S_i} &= \gamma_0 \overleftrightarrow{\chi J}_{ij} \vec{A}_j \end{aligned} \tag{20}$$

where

$$\begin{aligned}
\omega_1 &= -\dot{\beta} \sin\alpha + \dot{\gamma} \cos\alpha \sin\beta \\
\omega_2 &= \dot{\beta} \cos\alpha + \dot{\gamma} \sin\alpha \sin\beta \\
\omega_3 &= \dot{\alpha} + \dot{\gamma} \cos\beta \\
A_1{}^{\ell} &= \beta_\ell \sin\alpha + \gamma_\ell \cos\alpha \sin\beta \quad (\alpha_\ell = \frac{\partial\alpha}{\partial x_\ell} \text{ etc.}) \\
A_2{}^{\ell} &= \beta_\ell \cos\alpha + \gamma_\ell \sin\alpha \sin\beta \\
A_3{}^{\ell} &= \alpha_\ell + \gamma_\ell \cos\beta
\end{aligned}$$

and

$$\begin{aligned}
\chi_B &= N(0)\ (1-1/3Y(T))/[1-\bar{I}(1-1/3Y(T))] \\
(\chi^J)_{ij}^{\ell m} &= (N/m)Y(T)[\delta_{\ell m}\,\delta_{ij} - 3\langle k_\ell k_m f_i f_j\rangle] \\
Y(T) &= 2\pi T \sum_{n=0}^{\infty} \frac{\Delta^2}{(\omega_n^2+\Delta^2)^{3/2}} = \frac{(\rho s)}{\rho}\ \text{BCS}
\end{aligned} \tag{21}$$

and f_i is obtained from k_i by

$$(\vec{\sigma}\cdot\vec{f}) = R(\vec{\sigma}\cdot\vec{k})R^{-1} \tag{22}$$

$\vec{k}$ is the unit vector in the direction of the quasi-particle momentum $\vec{p}$ and finally $\langle A\rangle$ means the average of A over the direction of $\underline{p}$. We note here that the χ_B is the static susceptibility while $(\overleftrightarrow{\chi}^J)_{ij}$ is the static correlation function of the spin current in the B phase. Eq. (20) together with the spin conservation relation yield equations of motion for α, β and γ. This can be expressed in terms of the Lagrangian;

$$L = \int d^3r L(\vec{r})$$

and

$$L(\vec{r}) = \sum_{\alpha\beta\gamma} \dot{\alpha}\,\frac{E_{kin}}{\partial\dot{\alpha}} - E_{kin} - \chi_B\omega_L(\dot{\alpha} + \dot{\gamma}\cos\beta) - E_d \tag{23}$$

where E_{kin} is the kinetic energy given by

$$E_{kin} = 1/2\chi_B(\dot{\alpha}^2 + \dot{\beta}^2 + \dot{\gamma}^2 + 2\dot{\alpha}\dot{\gamma}\cos\beta) + 1/2\,(\chi^J)_{ij}^{\ell m} A_i^\ell A_j^m \tag{24}$$

and E_d is the dipole interaction energy.[6]

$$E_d(\alpha,\beta,\gamma) = \frac{2}{15}\chi_B\Omega_B^{\ 2}\{[(1+\cos\beta)(1+\cos\alpha_1) - \frac{3}{2}]^2 - \frac{5}{4}\} \tag{25}$$

and Ω_B is the longitudinal resonance frequency in the B phase and $\alpha_1 = \alpha + \gamma$. We note here that the first term in Eq. (24) is the same as the one for a spherical top, which follows from the isotropy of the static susceptibility in the BW state. Finally the equation of motion is given by

$$\frac{\partial}{\partial t}\left(\frac{\partial L}{\partial\dot{\alpha}}\right) + \frac{\partial}{\partial x_\ell}\left(\frac{\partial L}{\partial\alpha_\ell}\right) - \frac{\partial L}{\partial\alpha} = 0,\ \text{etc.} \tag{26}$$

which are the fundamental equations for the B phase. The spin wave dispersion is determined from small oscillation of the Eulerian angles around the equilibrium configuration. However, to do this we have first to determine the equilibrium configuration. Eq. (25) tells that the dipole interaction energy takes the minimum value whenever

$$(1+\cos\beta)\quad(1+\cos\alpha_1) = \frac{3}{2} \tag{27}$$

is satisfied. Since Eq. (27) defines a one dimensional manifold in the β-α_1 space (see Fig. 2), the dipole interaction energy is not adequate to determine the ground state configuration uniquely. In order to remove the remaining degeneracy we have to invoke other

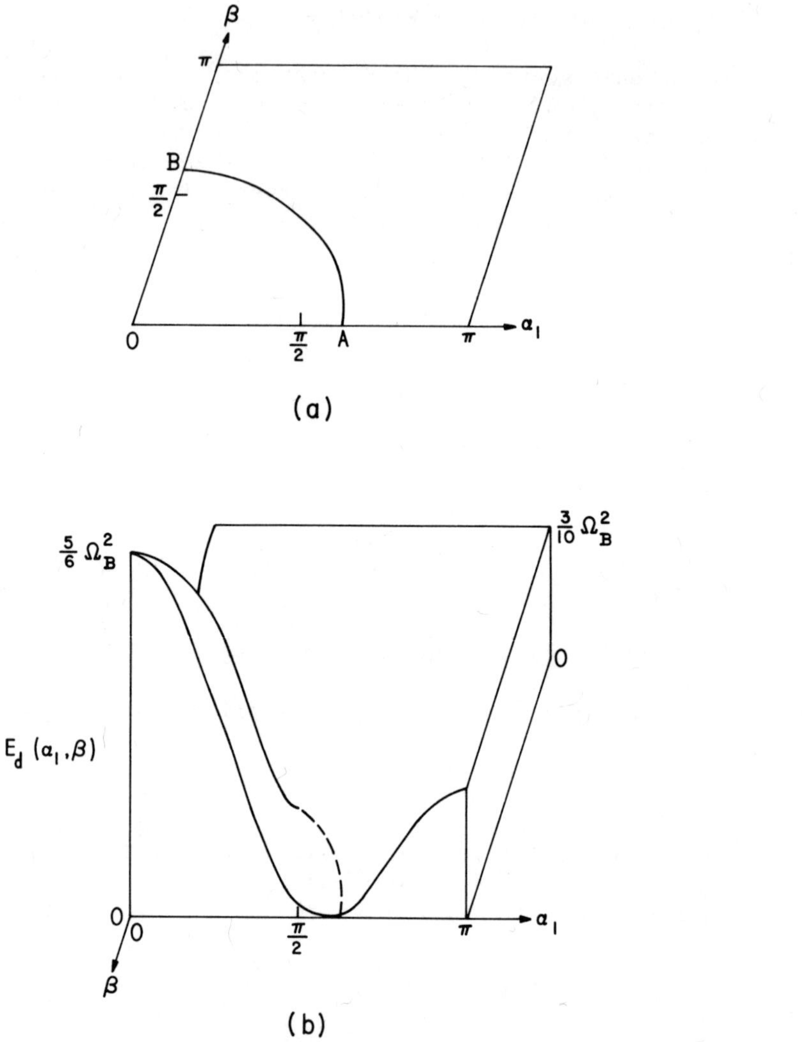

Figure 2. Dipole interaction energy $E_d(\alpha,\beta,\gamma)$ in the B phase shown as a function of β and $\alpha_1 = \alpha+\gamma$. (A quadrant is shown, since E_d is symmetric at $\beta=0$ and $\alpha+\gamma=0$.) The locus of the minimum in the dipole energy is shown by solid line in a). Points A and B in a) correspond with the two equilibrium configurations; the Leggett configuration and the wall fixed configuration.

possible perturbations, which have not been considered until now. The necessary perturbation energy appears to be provided either by the magnetic field[3] or by the wall of the vessel containing the liquid ^{3}He.[21] We will not go into details of these conditions, but limit ourselves to those two special configurations which are relevant in the experiment by Webb et al;[8,9]

a) Leggett Configuration: $[\alpha = \cos^{-1}(-1/4),\ \beta = \gamma = 0]$

This configuration is likely to be realized in a relatively strong magnetic field (along the z axis). If we keep the small deviations from the equilibrium configuration, we will have

$$\omega_1 = \dot{\phi}_1 = -\dot{\beta}\sin\alpha + \dot{\gamma}\,\beta\cos\alpha$$
$$\omega_2 = \dot{\phi}_2 = \dot{\beta}\cos\alpha + \dot{\gamma}\,\beta\sin\alpha$$
$$\omega_3 = \dot{\phi}_3 = \dot{\alpha} + \dot{\gamma} \tag{28}$$

and similar expressions for A_i^{ℓ}. Substituting Eq. (28) into the Lagrangian density (23), we obtain

$$\ddot{\phi}_1 = c^2[\nabla^2\phi_1 - 1/2\,\frac{\partial\psi_1}{\partial z}] - \Omega_B^{\ 2}\phi_1$$
$$\ddot{\phi}_2 = -\omega_L\dot{\phi}_3 + c^2[\nabla^2\phi_2 - 1/2\,\frac{\partial\psi_1}{\partial y}]$$
$$\ddot{\phi}_3 = \omega_L\dot{\phi}_2 + c^2[\nabla^2\phi_3 - 1/2\,\frac{\partial\psi_1}{\partial x}] \tag{29}$$

where

$$\psi_1 = \frac{\partial\phi_1}{\partial z} + \frac{\partial\phi_2}{\partial y} + \frac{\partial\phi_3}{\partial x}$$

and

$$c^2 = \frac{4}{15}\,v_F^{\ 2}[N(0)/\chi_B]Y(T) \tag{30}$$

Equation (29) implies that the spin wave modes couple with each other even in the case of small oscillations. When the spin wave is propagated along the z axis (parallel to the external field), these modes decouple into the longitudinal mode and the transverse mode with the dispersion

$$\omega^2 = (1/2)c^2q^2 + \Omega_B^{\ 2}$$

and

$$\omega\,(\omega \pm \omega_L) = c^2q^2 \tag{31}$$

respectively.

b) Wall Fixed Configuration: $[\alpha = \gamma = 0,\ \beta = \cos^{-1}(-1/4)]$

If the wall of the container consists of two plates parallel to the x-z plane, the present configuration will be realized in weak fields. In this case by introducing new variables

$$\phi_1 = \alpha+\gamma, \quad \phi_2 = \alpha-\gamma, \text{ and } \phi_3 = \beta-\cos^{-1}(-1/4) \tag{32}$$

we find

$$\ddot{\phi}_1 = -\sqrt{\frac{5}{3}}\,\omega_L\dot{\phi}_3 + c^2[\nabla^2\phi_1 - 1/8\ (\sqrt{15}\,\frac{\partial\psi_2}{\partial x} + \frac{\partial\psi_2}{\partial z})]$$

$$\ddot{\phi}_2 = \sqrt{\frac{5}{3}}\,\omega_L\dot{\phi}_3 + c^2[\nabla^2\phi_2 - 1/8\ (\sqrt{15}\,\frac{\partial\psi_2}{\partial x} + \frac{\partial\psi_2}{\partial z})]$$

and

$$\ddot{\phi}_3 = \sqrt{\frac{15}{4}}\,\omega_L(\dot{\phi}_1-\dot{\phi}_2) + c^2[\nabla^2\phi_3 - 1/2\,\frac{\partial\psi_2}{\partial y}] - \Omega_B^2\,\phi_3 \tag{33}$$

where

$$\psi_2 = \frac{\partial\phi_3}{\partial y} + \frac{3}{32}\,(\sqrt{15}\,\frac{\partial\phi_1}{\partial x} + \frac{\partial\phi_1}{\partial z}) + \frac{5}{32}\,(\sqrt{15}\,\frac{\partial\phi_2}{\partial x} + \frac{\partial\phi_2}{\partial z})\ . \tag{34}$$

In general, the spin wave dispersion is rather complicated. When the spin wave is propagated along the z axis, we will have

$$\omega^2 = \frac{31}{32}\,c^2q^2$$

and

$$\omega^2 = c^2q^2 + \frac{1}{2}\,\{\omega_L^2 + \Omega_B^2 \pm [(\omega_L^2 + \Omega_B^2)^2 + 4c^2q^2\omega_L^2]^{\frac{1}{2}}\} \tag{35}$$

for the longitudinal and the transverse wave respectively. These spin wave dispersions are consistent with the result due to Combescot.[13] We note that in this particular geometry the longitudinal mode is gapless.

V. RINGING OF MAGNETIZATION IN THE B PHASE

As another application of Eq. (26), we study the behavior of the magnetization[7] after a sudden application (or removal) of a magnetic field ΔH. It appears that the present theory can account for two distinct ringing behaviors in the B phase, as observed by Webb et al..[8] As before we will consider the ringing behavior of two configurations separately. We limit ourselves to the case $\Delta\vec{H}$ is parallel to $\vec{H}$.

a) Leggett configuration:

In the present situation (since $\beta=\gamma=0$ for all t), Eq. (26) reduces to

$$\ddot{\alpha} = \frac{16}{15}\,\Omega_B^2\;(\sin\alpha)\;(\cos\beta + 1/4) \tag{36}$$

Furthermore, the initial conditions are given by

$$\alpha = \cos^{-1}(-1/4),\ \text{and}\ \dot{\alpha} = \Delta\omega_L\ \text{at}\ t=0 \tag{37}$$

Then Eq. (36) is integrated to yield

$$\dot{\alpha}^2 = (\Delta\omega_L) - \frac{16}{15}\,\Omega_B^2\;(\cos\alpha + 1/4)^2 \tag{38}$$

and the magnetization is given by

$$M_z = \gamma_0\chi_B\dot{\alpha}\ ,\ M_x = M_y = 0\ . \tag{39}$$

Eq. (38) says that the (longitudinal) magnetization oscillates with ringing frequency ω_r;

$$\begin{aligned}\omega_r/\Omega_B &= \pi x\,[\xi^{1/2}kK(k)]^{-1}\ \text{for}\ 0<x<\sqrt{\tfrac{3}{5}}\\ &= \pi x\,[\xi^{1/2}K(k^{-1})]^{-1}\text{for}\ \sqrt{\tfrac{3}{5}}<x<\sqrt{\tfrac{5}{3}}\\ &= \pi x\,[\xi^{1/2}kK(k)]^{-1}\text{for}\ \sqrt{\tfrac{5}{3}}<x\end{aligned} \tag{40}$$

where

$$x = \Delta\omega_L/\Omega_B,\ \xi = (\sqrt{15}/4)x$$

and

$$k = 2\xi^{1/2}[(\xi+\tfrac{3}{4})\ (\xi+\tfrac{5}{4})]^{-1/2} \tag{41}$$

and K(k) is the complete elliptic integral. The ringing frequency given in Eq. (40) is calculated numerically and plotted in Fig. 3. For small $\Delta\omega_L$, $\omega_r = \Omega_B$ as expected. Then ω_r begins to decrease as $\Delta\omega_L$ increases and ω_r has a dip at $\Delta\omega_L = (\frac{5}{3})^{1/2}\Omega_B$. We expect another dip at $\Delta\omega_L = (\frac{5}{3})^{1/2}\Omega_B$. These dips correspond to two peaks in E_d along the line $\beta=0$ (at $\alpha_1=0$ and π) as seen in Fig. 2. The ringing frequency of M_z in the B phase has been recently measured by Webb _et al._[8,9] If we compare Ω_B with Ω_A, we have

$$\Omega_B/\Omega_A = (5/2)^{1/2}(\chi_A/\chi_B) \tag{42}$$

(we assumed here $\Delta_A^2 = \Delta_B^2$) as was first deduced by Leggett.[3] However,

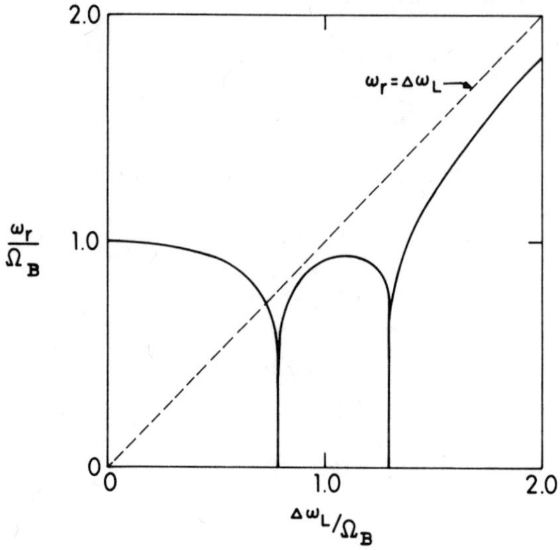

Figure 3. The ringing frequency ω_r for the Leggett configuration shown as a function of $\Delta\omega_L/\Omega_B$.

the observed ratio of these frequencies appears to be 1.9 ± 0.1. Furthermore, Webb et al.[8] observed that ω_r has a dip first around $\Delta\omega_L/\Omega_B \sim 1.5$, which is roughly twice larger than the present prediction.[B] The origin of these discrepancies is not understood at the moment.

b) Wall-fixed configuration ($\alpha=\gamma=0$, $\beta=\cos^{-1}(-1/4)$)

In this configuration we have to solve a set of equations (26) simultaneously. The initial conditions are given by

$$\omega_1 = 0, \quad \omega_2 = 0, \text{ and } \omega_3 = \Delta\omega_L \text{ at } t=0 \tag{43}$$

Introducing new variables by $\alpha_1 = \alpha+\gamma$, and $\gamma_1 = \alpha-\gamma$, we see easily that γ_1 is cyclic. Integration of the equation for γ_1 gives

$$\dot{\gamma}_1 = \frac{5}{4}(\Delta\omega_L-\omega_L)(1-\cos\beta)^{-1}+\omega_L \tag{44}$$

Substituting this into the expression of energy $\varepsilon = E_{kin} + E_d$, we have

$$\frac{1}{2}(\Delta\omega_L)^2 = \frac{1}{2}(1+\cos\beta)\dot{\alpha}_1^2 + \dot{\beta}^2 + \frac{1}{2}(1-\cos\beta)\left[\frac{\frac{5}{4}(\Delta\omega_L-\omega_L)}{1-\cos\beta}+\omega_L\right]^2 + \frac{2}{15}\Omega_B^2\left[(1+\cos\alpha_1)(1+\cos\beta)-\frac{3}{2}\right]^2 \tag{45}$$

In general we have to solve the equations containing two variables α_1 and β. However, if we confine ourselves to the case where ω_L, $|\Delta\omega_L| \ll \Omega_B$ as in the experiment by Webb et al.,[8,9] a further simplification is available.

In this limit the trajectory of the spin variables lies always very close to the locus of the minima of the dipole interaction energy (Eq. (27)). Introducing new variables by

$$\tan u = \cos\left(\frac{\beta}{2}\right) \left[\frac{1}{2}(1-\cos\alpha_1)\right]^{1/2}$$

and

$$\cos v = \frac{1}{2}(1+\cos\alpha_1)(1+\cos\beta) - 1 \tag{46}$$

with $v = \cos^{-1}(-\frac{1}{4})$, we can reduce Eq. (45) to

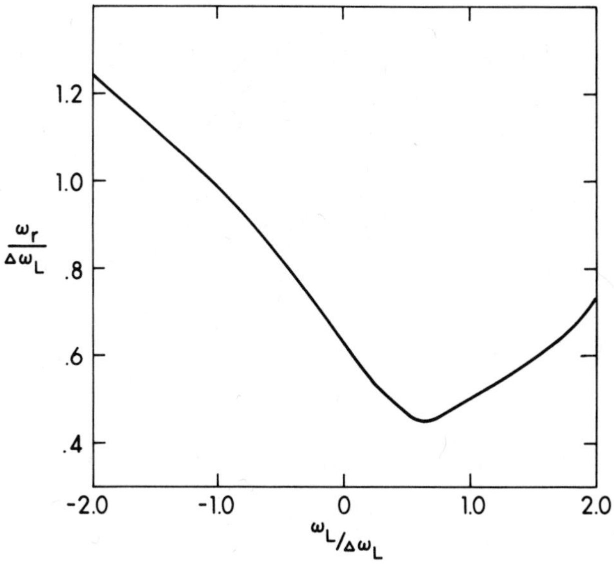

Figure 4. The ringing frequency ω_r for the Wall fixed configuration shown as a function of $\omega_L/\Delta\omega_L$ for $|\omega_L|$, $|\Delta\omega_L| \ll \Omega_B$.

$$\frac{1}{2}(\Delta\omega_L) = \frac{5}{4}\dot{u}^2 + \frac{5}{16}\cos^2 u\left(\frac{(\Delta\omega_L - \omega_L)^2}{\cos^2 u} + \omega_L\right)^2 \tag{47}$$

It is then easy to solve Eq. (47) in u; u oscillates with the fundamental frequency ω_r

$$\omega_r/\Delta\omega_L = \frac{\pi}{2\sqrt{2}K(k)}\left[\left(\frac{4}{5}-x\right)^2 + \frac{3}{5}x^2\right]^{1/4}$$

and

$$k^2 = \frac{1}{2}\left(1 - \frac{\frac{4}{5}-x}{\left[\left(\frac{4}{5}-x\right)^2 + \frac{3}{5}x^2\right]^{1/2}}\right) \tag{48}$$

where $x = \omega_L/\Delta\omega_L$ and K(k) is the complete elliptic integral. In this configuration ω_r is independent of Ω_B but proportional to $(\Delta\omega_L)$. The proportional factor depends on the ratio $\omega_L/\Delta\omega_L$, which is plotted in Fig. 4 for $2>x>-2$. Since the magnetization $\vec{M}$ can be

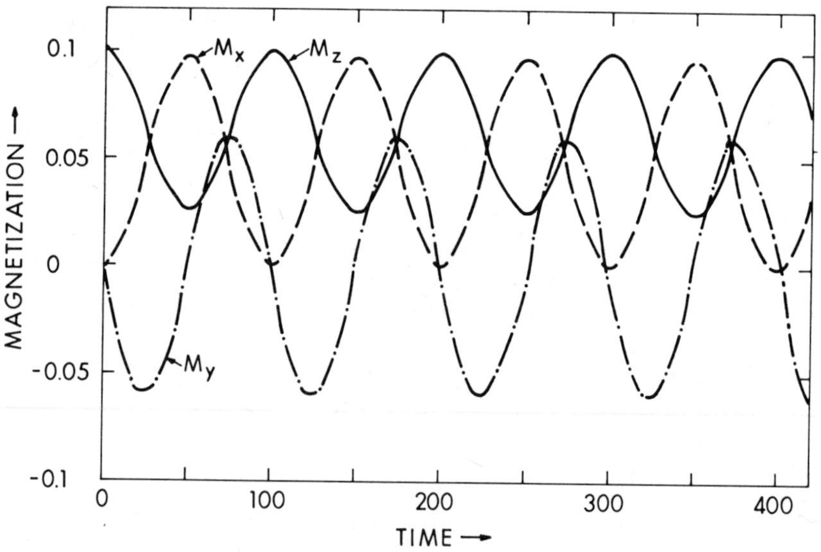

Figure 5. The time variation of the magnetization after a sudden removal of a magnetic field in the Wall fixed geometry shown as function of time for $\Delta\omega_L = 0.1\Omega_B$. The time is measured by Ω_B^{-1}.

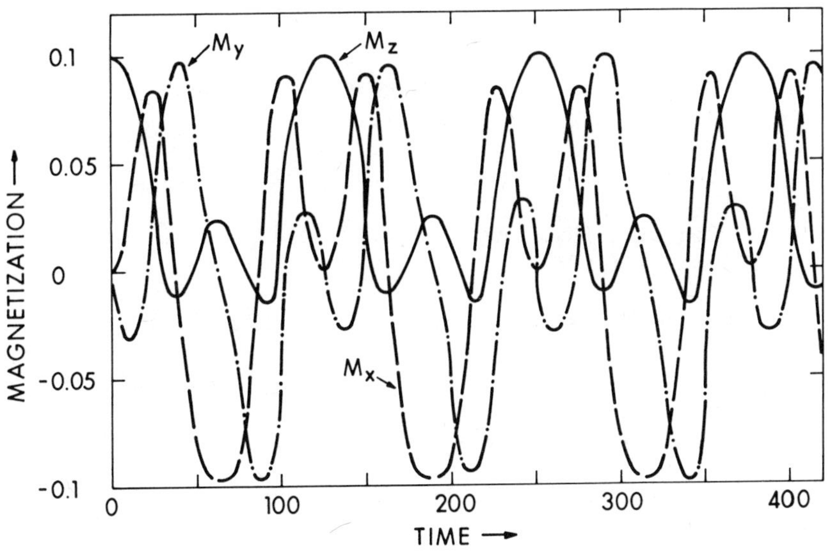

Figure 6. The time variation of the magnetization after a sudden application of a magnetic field. We took $\Delta\omega_L = \omega_L = 0.1\Omega_B$.

expressed in terms of u and $\dot{u}$, the above result predicts that $\vec{M}$ oscillates with the same frequency ω_r. Typical oscillations in the magnetization M_x, M_y, and M_z are plotted in Figs. 5 and 6 for $\Delta\omega_L = 0.1\ \Omega_B$, $\omega_L = 0$ and for $\Delta\omega_L = \omega_L = 0.1\ \Omega_B$ respectively. In contrast to the ringing discussed previously, in the present configuration the longitudinal ringing is always associated with the transverse ringing which should be easily measurable. The ringing frequency obtained here explains nicely the second ringing behavior in the B phase observed by Webb <u>et al</u>.[8,9] at relatively low temperature and in weak (or negligible) magnetic fields, although further experimental work in this regime is certainly desirable. We also note that the ringing frequency (Eq. (48)) for x=0 is obtained also by Brinkman,[22] based on a different equation (although it should be equivalent to ours) and with the additional boundary condition

$$\vec{S}\cdot\vec{n} = 0 \qquad (49)$$

where $\vec{n}$ is the axis of the instantaneous rotation of the spin space.

References

(1) D. D. Osheroff, W. J. Gully, R. C. Richardson, and D. M. Lee, Phys. Rev. Lett. 29 920 (1972).

(2) A. J. Leggett, Phys. Rev. Lett. 29 1227 (1972).

(3) A. J. Leggett, Phys. Rev. Lett. 31 352 (1973); and Ann. Phys. (N.Y.) 85 11 (1974).

(4) K. Maki and T. Tsuneto, Prog. Theor. Phys. 52 773 (1974).

(5) K. Maki and T. Tsuneto, Phys. Rev. B 11, Jan. (1975), in press.

(6) K. Maki, Phys. Rev. B 11, Feb. (1975), in press.

(7) K. Maki and C.-R. Hu, J. Low Temp. Phys. 18 377 (1975); and J. Low Temp. Phys. 19 259 (1975).

(8) R. A. Webb, R. L. Kleinberg, and J. C. Wheatley, Phys. Rev. Lett. 33 145 (1974).

(9) R. A. Webb, R. L. Kleinberg, and J. C. Wheatley, Phys. Lett. A 48 421 (1974).

(10) P. W. Anderson and W. F. Brinkman, Phys. Rev. Lett. 30 1108 (1973).

(11) V. Ambegaokar and N. D. Mermin, Phys. Rev. Lett. 30 81 (1973).

(12) K. Maki and H. Ebisawa, J. Low Temp. Phys. 15 213 (1974).

(13) R. Combescot, Phys. Rev. Lett. 33 946 (1974) and preprint.

(14) R. Combescot and H. Ebisawa, Phys. Rev. Lett. 33 811 (1974).

(15) K. Maki and H. Ebisawa, Phys. Rev. B 11, May (1975), in press.

(16) R. Combescot, Phys. Rev. Lett. 34 8 (1975).

(17) R. Balian and N. R. Werthamer, Phys. Rev. 131 1553 (1963).

(18) D. N. Paulson, J. Kojima, and J. C. Wheatley, Phys. Rev. Lett. 32 1098 (1974).

(19) A. I. Ahonen, M. T. Kaikala, and M. Krusius, and O. V. Lounasmaa, Phys. Rev. Lett. 33 628 (1974).

(20) K. Maki, J. Low Temp. Phys. 16 465 (1974).

(21) W. F. Brinkman, H. Smith, D. D. Osheroff, and E. I. Blount, Phys. Rev. Lett. 33 624 (1974).

(22) W. F. Brinkman, Phys. Lett. A 49 411 (1974).

DISSIPATIVE PROCESSES IN THE NUCLEAR MAGNETIC RESONANCES OF SUPERFLUID LIQUID ^{3}He*

Vinay Ambegaokar

Laboratory of Atomic and Solid State Physics
Cornell University
Ithaca, New York 14853

Abstract: A framework is given for the inclusion of relaxation processes in the theory of NMR in the superfluid phases of liquid ^{3}He, which allows one to compare two recent phenomenological theories. The effects of dissipation on large amplitude longitudinal ringing experiments are discussed.

Leggett's remarkable hydrodynamic theory[1] of NMR in the superfluid phases of liquid ^{3}He is based on two equations. The first, a conservation law, is the statement that the rate of change of the macroscopic spin angular momentum of the system is equal to the sum of the torques due to the Zeeman and dipole-dipole energies, the latter torque being anomalously large in a system with triplet spin pairing. The second equation is analogous to the Josephson condition in superconductivity and connects the motion of the spin degrees of freedom of the order parameter to the energy required for an infinitesimal increase in spin polarization.

With the substantial confirmation of the hydrodynamic theory, there is now increasing interest in the processes which lead to the damping of hydrodynamic oscillations. Combescot and Ebisawa[2] and, very recently, Leggett and Takagi[3] have proposed sets of equations including relaxation, which properly reduce to the hydrodynamic limit. Both theories can be used, among other things, to calculate the widths of resonance lines. The results are similar but differ

---

*Work supported by the National Science Foundation under grant number DMR-74-23494 and by the Materials Science Center, Cornell University, Technical Report #2425.

in detail.

The purpose of this note is to show how the two approaches are related, and to provide a conceptually simple framework for considering the problem. We show that the relaxation processes in question are contained in a certain correlation function, the detailed calculation of which requires microscopic considerations not given here. We argue on general grounds that the structure of this function is quite different in, on the one hand, normal and singlet-state paired systems, and, on the other, systems with triplet pairing. The discussion shows that the relaxation time introduced by Combescot and Ebisawa must approach infinity at the transition temperature T_c, whereas that of Leggett and Takagi need not, though it may well have a temperature dependence on the scale of T_c. We do not restrict ourselves to small deviations from the equilibrium orientation, and derive an equation, also obtained in Ref. 3 and by Combescot,[4] for large amplitude longitudinal ringing including damping. A qualitative explanation for the experimental deviations[5] from simple theory for this phenomenon is thereby suggested, which explanation can be quantitatively tested by further experiments.

It is convenient, following Combescot,[6] to work in a reference frame tied to the spin axes of the order parameter.[7] Let the continuous time-dependent unitary operator which generates the transformation from the laboratory frame be $e^{-i\vec{S}\cdot\vec{\theta}(t)}$ where $\vec{S}$ is the total spin angular momentum operator of the many-body system. The transformed Hamiltonian of the system is then

$$\mathcal{H}' = e^{-i\vec{S}\cdot\vec{\theta}}\mathcal{H}e^{i\vec{S}\cdot\vec{\theta}} - i\, e^{-i\vec{S}\cdot\vec{\theta}}\frac{\partial}{\partial t}e^{i\vec{S}\cdot\vec{\theta}} \,. \tag{1}$$

The last term may be reduced using the commutation relations of the spin to the form $\vec{S}\cdot[\dot{\theta}\,\hat{\theta} + \sin\theta\,\dot{\hat{\theta}} + (1 - \cos\theta)\,\theta \times \dot{\hat{\theta}}] \equiv \vec{S}\cdot\vec{C}_{\theta}$. Adding this form to the transformed Zeeman energy (γ-nuclear moment; $\vec{H}$-magnetic field), $-\gamma\vec{S}\cdot R_{\vec{\theta}}\vec{H}$, where $R_{\vec{\theta}}$ is the 3×3 matrix corresponding to the rotation $\vec{\theta}(t)$, one obtains the effective magnetic energy ($\gamma\vec{H} \equiv \vec{\omega}_L$)

$$\mathcal{H}'_z(t) = -\vec{S}\cdot[R_{\vec{\theta}(t)}\,\vec{\omega}_L - \vec{C}_{\vec{\theta}(t)}] \,. \tag{2}$$

[We remark in passing that the object $\vec{C}_{\vec{\theta}}$ may be interpreted usefully in two ways. It describes precession: any vector $\vec{T}$ fixed in the rotating frame can be shown to satisfy the equation $d\vec{T}/dt = \vec{T} \times R^{-1}_{\vec{\theta}(t)}\,\vec{C}_{\vec{\theta}(t)}$. Additionally, and characteristically of off-diagonal order, $R^{-1}_{\vec{\theta}}\vec{C}_{\vec{\theta}}$ ($= -\vec{C}_{-\vec{\theta}}$) determines the rate of change of the order parameter. This quantity transforms under (1) according to $\Delta_{\alpha\beta}(\hat{k}t) = [\exp i \frac{\vec{\sigma}}{2} \cdot \vec{\theta}(t)]_{\alpha\gamma} [\exp i \frac{\vec{\sigma}}{2} \cdot \vec{\theta}(t)]_{\beta\delta} \Delta'_{\gamma\delta}(\hat{k})$ and thus obeys the equation $i \, \partial \Delta_{\alpha\beta}/\partial t = \vec{C}_{-\vec{\theta}} \cdot [\frac{1}{2} \vec{\sigma}_{\alpha\gamma} \delta_{\beta\delta} + \frac{1}{2} \delta_{\alpha\gamma} \vec{\sigma}_{\beta\delta}] \Delta_{\gamma\delta}$, assuming that Δ' is independent of time.]

Suppose we now calculate the linear response[8] of the average spin of the system to the perturbation (2), imagined to be turned on adiabatically. The result will have the form

$$\Delta[R_{\vec{\theta}} \langle \vec{S} \rangle] = \frac{1}{\gamma^2} \int_0^t dt' \, \vec{\chi}'(t, t', 0) \cdot [R_{\vec{\theta}(t')} \vec{\omega}_L - \vec{C}_{\vec{\theta}(t')}] \, . \quad (3)$$

We shall see that this is the generalization of Leggett's second equation (Ref. 3, Eq. (2)) to include relaxation. $\vec{\chi}'$ is the time-dependent susceptibility of the system in the rotating frame. In several situations $\vec{\chi}'$ can be easily calculated. In the normal state, with or without spin-conserving collisions, the small dipole energy and other spin-changing interactions work on a time scale longer than any of interest in resonance phenomena, and $\vec{\chi}'$ is determined by the equal time commutators of the spin variables. One thus finds

$$(\chi_N')_{ij} = -\gamma^2 \varepsilon_{ijk} [R_{\vec{\theta}(0)} \langle \vec{S}(0) \rangle]_k \quad . \quad (4)$$

In this case the physics of Eq. (3) becomes obvious by letting $t \to 0$, whereupon (3) and (4) simply yield the equation of precession $\Delta \langle \vec{S} \rangle / \Delta t = \langle \vec{S} \rangle \times \vec{\omega}_L$ viewed in an arbitrary rotating coordinate system, there now being no special meaning to the rotation $\vec{\theta}(t)$. For the same reasons, Eq. (4) also holds for a system with singlet-state pairing, again with or without spin-conserving collisions.

For a system with triplet pairing, the calculation of χ' is only trivial in the collisionless limit. In this case one finds that, at fixed orientation of the gap anisotropy, a part of the magnetization of the system follows a magnetic field on a time scale $(\hbar/\Delta)$, where Δ is the energy gap. Except extremely close to T_c, this time is small compared to resonance periods and lifetimes, and (3) becomes a relationship local in time

$$R_{\vec{\theta}(t)} \langle \vec{S}(t) \rangle = \frac{1}{\gamma^2} \vec{\chi}_1' \cdot [R_{\vec{\theta}(t)} \vec{\omega}_L - \vec{C}_{\vec{\theta}(t)}] \, . \quad (5)$$

In simple pairing theory (no Fermi liquid effects) one finds

$$(\chi_1')^0_{ij} = -\frac{\gamma^2}{2} \sum_{\vec{k}} \frac{\Delta_i' \Delta_j'^* + \Delta_j' \Delta_i'^* - 2\delta_{ij} |\Delta'|^2}{4E'^2} \frac{1 - 2f(E')}{E'} \quad (6)$$

where $\Delta'_i(\hat{k})$ is the vector describing the triplet state order $[\Delta'_{\alpha\beta}(\hat{k}) = i(\vec{\Delta}'\cdot\vec{\sigma})\sigma_2|_{\alpha\beta}]$ in the rotating frame; $E'^2 = \varepsilon^2 + |\Delta'|^2$; and f is the Fermi function. Fermi liquid effects can be approximately included in (5) in the usual way by the replacements $\vec{\chi}_1' \rightarrow (\vec{\chi}_1')^0$ and $\vec{\omega}_L \rightarrow \vec{\omega}_L - \gamma^2 f_o^a \vec{S}$, f_o^a being the spin-antisymmetric, isotropic Landau parameter.

A few words about the dipole energy are needed here. Since this interaction, which is in principle included in (6) through the self-consistent gap equation determining $\vec{\Delta}'$, is not invariant under the transformation (1), $\vec{\Delta}'$ will have an amplitude fluctuation as the spin axes rotate with respect to the orbital ones. These amplitude fluctuations are, however, of relative order $N(0)\gamma^2 \sim 10^{-7}$ (where $N(0)$ is the density of states at the Fermi surface) and may be neglected in most situations. This approximation is identical to one made by Leggett[1] at a similar stage of his calculation, and it makes χ_1' independent of time. One may transform to the laboratory frame by defining $\vec{\chi}_1 = R^{-1}\,\vec{\chi}_1'\,R$. Eq. (5) is then seen to imply, via the discussion following Eq. (2), that in the collisionless limit the spin-anisotropy of the order parameter precesses about $\vec{\omega}_L - \gamma^2\,\vec{\chi}_1^{\,-1}\langle\vec{S}\rangle$, or, equivalently, that the rate of change of the order parameter is determined by the energy $-\vec{s}\cdot(\vec{\omega}_L - \gamma^2\,\vec{\chi}_1^{\,-1}\langle\vec{S}\rangle)$, where $\vec{s}$ is the spin of a pair. Viewed in this second way, Eq. (5) is seen to be the spin-triplet generalization of a Josephson equation (c.f. Ref. 1).

It is important to note that Eq. (5) is not in conflict with spin conservation, even when the dipole energy is neglected in χ_1.[6,7] The dipole energy enters the conservation law (Leggett's first equation)

$$\frac{\partial\langle\vec{S}\rangle}{\partial t} - \langle\vec{S}\rangle \times \vec{\omega}_L + \frac{\partial}{\partial\vec{\theta}} E_{dip}\left\{R_{\vec{\theta}}^{-1}\,\vec{\Delta}'\right\} \qquad (7)$$

and (7) and (5) together determine the driving force for the susceptibility at fixed orientation. In particular, if one ignored the dipole energy in (7), the magnitude of $\langle\vec{S}\rangle$ would be constant. For the special case of $\langle\vec{S}\rangle = 0$, Eq. (5) would then state that the anisotropy axes precess about $\vec{\omega}_L$, i.e., that the rate of change of the order parameter is determined by the magnetic field alone.

Leggett's second hydrodynamic equation is, in the present language, Eq. (5) with $\vec{\chi}_1'$ replaced by $\vec{\chi}'$ (the thermodynamic susceptibility at fixed orientation),[9] he having intuited that this would be the effect of spin-conserving collisions acting on a time scale small compared to a resonance period.

The theories of both Ref. 2 and 3 interpolate between the collisionless and the collision-dominated limits. To avoid complications irrelevant for the purposes of comparing these theories, we consider the longitudinal case in which $\langle\vec{S}\rangle$, $\vec{\theta}$, and $\vec{\omega}_L$ are all in the same direction (z). By giving the single particle excitations a phenomenological width and ignoring "scattering-in" terms (or, equivalently, vertex corrections) one obtains Combescot and Ebisawa's result for the zz component of the response function[10]

$$\chi_{zz}(t,t') = (\chi_1)_{zz}\,\delta(t-t') + e^{-\frac{t-t'}{\tau_{CE}}}\,\frac{1}{\tau_{CE}}\,(\chi-\chi_1)_{zz}\,. \tag{8}$$

We see at once that the correct normal state limit ($\chi_{zz} = 0$) requires $\tau_{CE}(T \to T_c) \to \infty$, which calls into question the experimental fit in Ref. 2 from which the relaxation time at T_c was inferred to be of the order of the normal state spin-diffusion lifetime.

Leggett and Takagi,[3] as a result of an instructive thought experiment, write a phenomenological rate equation for the collision-dragged part of the spin ($\vec{\tilde{S}}_q$) as it comes into equilibrium with the total spin $\vec{S}$, under the influence of spin-conserving collisions. The z-component of this equation is

$$\frac{\partial \tilde{S}_{qz}}{\partial t} + \frac{1}{\tau_{LT}}\,\tilde{S}_{qz} = \frac{(\chi - \chi_1)\,\chi^{-1}}{\tau_{LT}}\,S_z\,. \tag{9}$$

If, following them, we interpret S_z as $\tilde{S}_{pz} + \tilde{S}_{qz}$ where $\tilde{S}_{pz}$ (the spin-polarization due to superflow) is given by (5), we find that (9) can be written as a rate equation for $\tilde{S}_q$ to approach $(\chi - \chi_1)\,\chi_1^{-1}\tilde{S}_p$ from which follows the longitudinal response function[11]

$$\chi_{zz}(t,t') = (\chi_1)_{zz}\delta(t-t') + e^{-\frac{\chi_1\,\chi^{-1}}{\tau_{LT}}(t-t')}\,\frac{(\chi-\chi_1)}{\tau_{LT}}\,\chi^{-1}\chi_1\Big|_{zz}. \tag{10}$$

Eq. (8) and (10) show the connection between the two theories. Since $\chi_1(T_c) = 0$, Eq. (10) gives the correct normal state limit with $\tau_{LT}(T_c)$ finite. It seems likely, however, that coherence factors and the details of the collisional process will make the width analogous to $(\tau_{LT})^{-1}$ in the microscopic theory temperature dependent on the scale of T_c, as well as anisotropic.

All of the published results[2,3] on CW resonance linewidths follow easily from Eq. (3), (7), and (8) or (10). For the transverse resonances in the experimentally interesting region $\omega_L\,\tau_{CE} \ll 1$, Eq. (8) and (10) retain the form given, except that the tensor expressions for χ and χ_1 must be used.

We remark finally that from (3), (7), and (10) there immediately follows the equation for longitudinal ringing:[3,4]

$$\tau_{LT}\dddot{\theta} + \ddot{\theta} + \tau_{LT}\frac{\gamma^2}{\chi_1}\frac{\partial^2 E_{dip}}{\partial\theta^2}\dot{\theta} + \frac{\gamma^2}{\chi}\frac{\partial E_{dip}}{\partial\theta} = 0 \quad . \tag{11}$$

One consequence of this equation is that the damping terms, though small, can have a strong cumulative effect on low frequency hydrodynamic modes. Thus, one should find very interesting histories: fast rotation → slow rotation → slow vibration → fast vibration. In particular, the observation[5] of faster vibrations than those predicted by the hydrodynamic theory (i.e., the second and fourth terms of Eq. (11)) could be due to a time lapse before the oscillations are first observed. Further consequences of Eq. (11) will be published elsewhere with M. Levy.

I have benefited greatly from conversations with R. Combescot, N. D. Mermin, D. Rainer, T. L. Ho, and M. Levy. I am very grateful to A. J. Leggett and S. Takagi for a prepublication copy of their work.

References

(1) A. J. Leggett, Ann. Phys. (N.Y.) 85, 11 (1974).

(2) R. Combescot and H. Ebisawa, Phys. Rev. Lett. 33, 810 (1974).

(3) A. J. Leggett and S. Takagi, to be published.

(4) R. Combescot, to be published.

(5) D. N. Paulson, H. Kojima, and J. C. Wheatley, Phys. Lett. 47A, 457 (1974).

(6) R. Combescot, Phys. Rev. A10, 1700 (1974).

(7) c.f., V. Ambegaokar and L. P. Kadanoff, Nuovo Cimento 22, 914 (1961) for the analogous way of dealing with phase (or density) oscillations in superconductors.

(8) A linear calculation suffices because the effective magnetic energy per particle is always much less than the Fermi energy.

(9) The pairing theory expression for $\overleftrightarrow{\chi}'$ is given, with obvious minor differences in notation, in R. Balian and N. R. Werthamer, Phys.

Rev. 131, 1553 (1963), Eq. (49). Fermi liquid effects may then be approximately included by the procedure described below Eq. (6).

(10) To be precise, the lifetime τ introduced in Ref. 2 differs from τ_{CE} because our quantity contains Fermi liquid corrections. The exact correspondence is $\tau_{CE} = \tau(1 + f_o^a\chi_1^o)(1 + f_o^a\chi^o)^{-1}$.

(11) Although Eq. (9) gives precisely the results of Ref. 3 with τ_{LT} identified with the lifetime introduced there, there is a slight difference in point of view. In Ref. 3 Fermi liquid effects would not be included on the right side of our Eq. (9), but the equation corresponding to our (5) is modified in a compensating way, with the result that our Eq. (5) and (9) are entirely equivalent to, and, because Fermi liquid corrections are tucked out of sight, seem slightly simpler than, Eq. (7) and (8) of Ref. 3.

HYDROMAGNETIC EFFECTS IN SUPERFLUID $^3He^*$

Alexander L. Fetter

Institute of Theoretical Physics, Department of Physics,
Stanford University
Stanford, California 94305

Abstract: A Ginzburg-Landau expansion is used to study a uniformly-flowing p-wave Fermi superfluid in a static magnetic field near T_c. The three-dimensional state undergoes strong hydromagnetic suppression, affecting the critical current, the dipole energy, and the overall phase diagram.

The new low-temperature phases of ^3He have attracted great interest owing to their similarity both to superfluid ^4He and to metallic superconductors. In particular, these new states are thought to contain coherent spin-triplet Cooper pairs of ^3He atoms with unit orbital angular momentum. The macroscopic quantum condensate tends to control the remaining particles, leading to a two-fluid hydrodynamics similar to that familiar in ^4He. On the other hand, the free energy and thermal properties should be analogous to those of metallic superconductors, and a "Ginzburg-Landau" expansion might be expected to hold near the transition temperature. One significant difference, however, is the absence of electrical charge, so that ^3He displays neither a Meissner effect nor orbital magnetism. Consequently, superflow in ^3He should be uniform instead of exponentially screened. This charge neutrality renders the spin magnetism accessible to experiment, and such studies have provided much of the available information.[1]

The detailed description relies on a gap function $\Delta_{\alpha\beta}$ generalized to a 2×2 matrix in spin space, with $\alpha\beta$ referring to spin indices

---

*Research sponsored in part by Air Force Office of Scientific Research grant AFOSR-74-2576.

↑ and ↓.[2] Since the Cooper pairs are spin triplets, $\Delta_{\alpha\beta}$ is symmetric and hence expressible as a linear combination of the unit matrix and the two Pauli matrices σ_1 and σ_3. Equivalently, it is conveniently written $\Delta_{\alpha\beta} = \Delta_\mu(i\sigma_\mu\sigma_2)_{\alpha\beta}$, where Δ_μ transforms like a vector under spin rotations. For p-wave pairs, $\Delta_{\alpha\beta}$ must also transform like a vector under spatial rotations and may be written as a linear function of the unit vector $\hat{k}$ specifying the direction on the Fermi surface $\Delta_\mu(\hat{k}) = A_{\mu i}\hat{k}_i$. Here the tensor $A_{\mu i}$ has the dimensions of energy and its indices μi refer to spin and spatial vectors. (We here follow the convention of Mermin and Stare,[3] which unfortunately differs from that of Brinkman, Serene, and Anderson.[4] Also, repeated indices are summed from 1 to 3 unless otherwise noted.) In the simplest case of stationary pairs, the gap parameter is a spatial constant. More generally, $A_{\mu i}$ may vary slowly with position; for example, states with uniform flow have a phase factor

$$A_{\mu i}(\mathbf{r}) = e^{i\mathbf{q}\cdot\mathbf{r}}A_{\mu i} \; . \tag{1}$$

Consider first the uniform static configuration with no magnetic field. The corresponding free-energy density near the transition temperature T_c may be expanded in the general form[3,4]

$$\begin{aligned} F_0 = {} & -\alpha A_{\mu i}{}^* A_{\mu i} + \beta_1 A_{\mu i}A_{\mu i}A_{\nu j}{}^* A_{\nu j}{}^* + \beta_2 A_{\mu i}{}^* A_{\mu i}A_{\nu j}{}^* A_{\nu j} \\ & + \beta_3 A_{\mu i}{}^* A_{\nu i}{}^* A_{\nu j}A_{\mu j} + \beta_4 A_{\mu i}{}^* A_{\nu i}A_{\nu j}{}^* A_{\mu j} + \beta_5 A_{\mu i}{}^* A_{\nu i}A_{\nu j}A_{\mu j}{}^* \end{aligned} \tag{2}$$

where α is a positive constant with dimensions of (energy×volume)$^{-1}$ that vanishes as $T_c - T \to 0^+$. The simplest weak-coupling model takes a nonlocal spin-independent potential of the form $(V\mathbf{x},\mathbf{x}') = -3(g/k_F{}^2)(\nabla\cdot\nabla')\delta(\mathbf{x})\delta(\mathbf{x}')$, and a detailed calculation for $N(0)g \ll 1$ yields[5] $\alpha = \frac{1}{3}N(0)(1 - T/T_c)$, with $N(0) = mk_F/2\pi^2\hbar^2$ the density of states of one spin projection. Similarly the β's are of order $N(0)(k_BT_c)^{-2}$, with detailed numerical values that depend sensitively on the effective interactions; in the weak-coupling limit, they become[5]

$$-2\beta_1 = \beta_2 = \beta_3 = \beta_4 = -\beta_5 = 7\zeta(3)(120\pi^2)^{-1}N(0)(k_BT_c)^{-2}. \tag{3}$$

The introduction of hydrodynamic flow adds a term F_K representing the kinetic energy of the particles.[6] For weak coupling near T_c, the form of F_K is most readily determined from an explicit calculation of the current density to second order in the gap function,[7,8] which yields

$$J_i = 4K\hbar^{-1}\mathrm{Im}(A_{\mu i}{}^*\partial_j A_{\mu j} + A_{\mu j}{}^*\partial_i A_{\mu j} + A_{\mu j}{}^*\partial_j A_{\mu i}) \tag{4}$$

where $K = 7\zeta(3)(240\pi^2)^{-1}N(0)(\hbar v_F/k_BT_c)^2 = \frac{1}{5}N(0)\xi_0^2$, and $\xi_0 = [7\zeta(3)/48\pi^2]^{1/2}(\hbar v_F/k_BT_c) \approx 120$ Å for ^{3}He. De Gennes' procedure[6] of ascribing an infinitesimal test charge 2e to each Cooper pair then implies

$$F_K = K(\partial_i A_{\mu i}{}^* \partial_j A_{\mu j} + \partial_i A_{\mu j}{}^* \partial_i A_{\mu j} + \partial_i A_{\mu j}{}^* \partial_j A_{\mu i}), \tag{5}$$

which differs somewhat from that originally proposed.[6] In addition, application of a magnetic field augments the free-energy density by an amount $-\frac{1}{2}\chi_{\mu\nu}H_\mu H_\nu$.[9] Here $\chi_{\mu\nu}$ is the positive susceptibility tensor; it decreases by an amount of order $\chi_n(\Delta/k_BT_c)^2$ on entering the condensed state, and the free energy thus acquires a positive term[5,10]

$$F_Z = g_Z H_\mu A_{\mu i}{}^* A_{\nu i} H_\nu \tag{6}$$

where $g_Z = 7\zeta(3)(48\pi^2)^{-1}N(0)(\gamma\hbar/k_BT_c)^2$, and $\frac{1}{2}\gamma\hbar\underset{\sim}{\sigma}$ is the nuclear magnetic moment of a ^{3}He atom ($\gamma\hbar \approx -2.1 \times 10^{-23}$ erg/gauss). The final contribution to the energy density arises from the magnetic dipole-dipole interaction. The calculation starts from the first-quantized total dipole energy and eventually yields the approximate form[11]

$$F_D = g_D(A_{\lambda\lambda}{}^* A_{\mu\mu} + A_{\mu i}{}^* A_{i\mu} - \frac{2}{3} A_{\mu i}{}^* A_{\mu i}) \tag{7}$$

where $g_D = (\pi/10)[N(0)\gamma\hbar\ln(1.13\hbar\omega_0/k_BT_c)]^2$ and $\hbar\omega_0/k_B$ is a cut-off temperature of order 1K.

The equation for $A_{\mu i}$ now follows by requiring the total free energy to be stationary under variations of $A_{\mu i}{}^*$, plus appropriate boundary conditions.[6] For simplicity, we shall treat only an unbounded uniform medium, when surface effects are negligible, and $A_{\mu i}(\underset{\sim}{r})$ is given by Eq. (1). The kinetic energy then reduces to

$$F_K = 15K(4\pi)^{-1} \int d\Omega_k (\hat{k}\cdot\underset{\sim}{q})^2 \, |\underset{\sim}{\Delta}(\hat{k})|^2 \quad . \tag{8}$$

Moreover, the usual choice of unitary states identifies $|\underset{\sim}{\Delta}(\hat{k})|^2$ as the squared energy gap in the direction $\hat{k}$, implying that the gap should be oriented with its minimum values along the direction of hydrodynamic flow. Similarly, a simple rearrangement of Eq. (6) gives

$$F_Z = 3g_Z(4\pi)^{-1} \int d\Omega_k |\underset{\sim}{H}\cdot\underset{\sim}{\Delta}(\hat{k})|^2 \quad , \tag{9}$$

implying that the vector $\underset{\sim}{\Delta}(\hat{k})$ should if possible lie perpendicular to $\underset{\sim}{H}$. Thus $\underset{\sim}{H}$ affects the direction of the spin vector $\underset{\sim}{\Delta}$, whereas $\underset{\sim}{q}$ affects the angular orientation of the squared energy gap. In contrast, the dipole energy couples the spin and spatial axes through the first two terms of (7), orienting these otherwise independent systems.

It is not difficult to obtain the nonlinear differential equation for $A_{\mu i}(\underset{\sim}{r})$ by varying $\int d^3r(F_0 + F_K + F_Z + F_D)$, but the presence of the dipole contribution greatly complicates the final equation. Fortunately, F_D is a small correction except near T_c, where $\alpha \lesssim g_D$, or, equivalently where $1 - T/T_c \lesssim N(0)(\gamma\hbar)^2[\ell n(1.13\hbar\omega_0/k_BT_c)]^2 \approx 2\times10^{-6}$. Consequently, the dipole terms may be omitted in determining $A_{\mu i}$, which then contains undetermined parameters specifying the relative orientation of the spin and spatial vectors. If this residual freedom is used to minimize the dipole energy, the total free energy will be correct through order g_D, although the actual $A_{\mu i}$ need not be so accurate.

To introduce this approach, we recall the known unitary p-wave states in the absence of flow (q=0) or magnetic field (H=0).[12,13] Four distinct choices yield local minima of the free energy $\int d^3rF_0$: The axial (ABM) states have the intrinsically complex form

$$A_{\mu i} = \Delta\hat{v}_\mu(\hat{n}^1 + i\hat{n}^2)_i \tag{10}$$

with $\hat{v}$, $\hat{n}^1$, and $\hat{n}^2$ arbitrary unit vectors, apart from the restriction $\hat{n}^1 \cdot \hat{n}^2 = 0$. For this state, the vector $\hat{\ell} = \hat{n}^1 \times \hat{n}^2$ represents the direction of orbital angular momentum of the Cooper pair, wholly analogous to the circular polarization state of the (spin 1) photon. The other three possible states may be chosen with intrinsically real $A_{\mu i}$ of the form

$$A_{\mu i} = \Delta \sum_{=1}^{p} \hat{v}_\mu^\lambda \hat{n}_i^\lambda \tag{11}$$

where $\hat{v}^\lambda$ and $\hat{n}^\lambda$ are unit vectors obeying the restrictions $\hat{v}^\kappa \cdot \hat{v}^\lambda = \delta_{\kappa\lambda} = \hat{n}^\kappa \cdot \hat{n}^\lambda$ and p = 1,2, or 3 denotes the dimensionality. These states are also called polar, planar, and isotropic (or BW). The corresponding gap parameter and free energy become

$$|\Delta|^2 = \alpha/2p\bar{\beta}, \quad F_0 = -\alpha^2/4\bar{\beta} \tag{12}$$

where $\bar{\beta} = \beta_{245}$ and p = 2 for the axial state, and $\bar{\beta} = \beta_{12} + p^{-1}\beta_{345}$ for the remaining three; here and henceforth, we use β_{ij} and β_{ijk} to denote the sum of the appropriate β's.[13] An equivalent and more useful form for the three-dimensional state is

$$A_{\mu i} = \Delta R_{\mu i} \tag{13}$$

with R an arbitrary three-dimensional rotation matrix.

If β_{13} and β_{345} are both positive (as seems to describe the experimental situation), the axial state is the true minimum for $\frac{3}{2}\beta_{13} > \beta_{345} > \beta_{13} - \beta_{12}$, whereas the three-dimensional state is the true minimum for $\beta_{345} > \frac{3}{2}\beta_{13}$. In addition, the two-dimensional state has lower free energy than the axial state if $\beta_{345} > 2\beta_{13}$. Comparison with Eq. (3) shows that the weak-coupling theory predicts a three-dimensional equilibrium state, with the two-dimensional and axial states degenerate and lying somewhat higher. The experimental β's are thought to approach the weak-coupling values at low pressure, identifying the observed B phase as a three-dimensional one. In contrast, the A phase (assumed axial) arises from pressure-dependent contributions to the β's that cause $\beta_{345} - \frac{3}{2}\beta_{13}$ to vanish at the polycritical point and along the observed AB transition line. Note that the β's also depend weakly on temperature at fixed pressure, unlike the usual situation in the Landau theory.

We now consider the effect of relatively strong magnetic fields and flows, characterized by the conditions $g_Z H^2$ and Kq^2 both $\gg g_D$ (i.e. $H \gtrsim 100$ gauss and $q \gtrsim 1.4 \times 10^3\ \mathrm{cm}^{-1}$; alternatively, the flow velocity $v = \hbar q/2m_3$ must exceed ≈ 0.15 cm sec^{-1}). When these restrictions are satisfied, the order parameter [Eq. (1)] satisfies the nonlinear field equations with no dipole contribution. Equations (8) and (9) suggest that the spin vectors $\hat{v}$ should lie perpendicular to $\underset{\sim}{H}$ and the orbital vectors $\hat{n}$ perpendicular to $\underset{\sim}{q}$. This configuration is readily achieved for the axial state and for the one- and two-dimensional ones, but not for the three-dimensional one, which must be treated separately. In particular, the one-dimensional state orients the symmetry axis of its squared energy gap along $\hat{n}$ with angular dependence $(\hat{k}\cdot\hat{n})^2$; on the other hand, the axial and two-dimensional states orient the symmetry axis of $|\Delta(\hat{k})|^2$ along $\underset{\sim}{q}$ with angular dependence $1 - (\hat{k}\cdot\hat{q})^2$. In addition, the field equations for these three states yield a reduced gap coefficient and free energy, obtained with the substitution $\alpha \to \alpha - Kq^2$ in Eq. (12). The magnetic field affects neither of these physical quantities, implying that the corresponding static magnetic susceptibility remains at the normal value, and the hydrodynamic flow merely scales the overall free energy of these three states, leaving their relative positions unaltered.

Given the matrix $A_{\mu i}$, Eq. (4) immediately provides the mass current for these three states

$$\underset{\sim}{J} = 2K\hbar^{-1}\underset{\sim}{q}(\alpha - Kq^2)\bar{\beta}^{-1} \tag{14}$$

which is unaffected by the magnetic field. Its dependence on q is wholly analogous to that for a superconducting film,[14] initially increasing linearly with q, but reaching a maximum value $J_c = (4\alpha/3\bar{\beta}\hbar)(\alpha K/3)^{1/2}$ at a critical wave number $q_c = (\alpha/3K)^{1/2}$, above which the states simply cease to exist. This critical current is proportional to $(T_c - T)^{3/2}$ and such a phenomenon might be observable near T_c assuming that no other mechanism (like vortex nucleation) intervenes. The corresponding critical velocity $hq_c/2m_3$ varies like $(T_c - T)^{1/2}$.

Before treating the three-dimensional state, we shall briefly consider the dipole energy, which tends to orient the $\hat{v}$'s perpendicular to the $\hat{n}$'s. In the one-dimensional state, the previous restrictions $\hat{v}\perp H$ and $\hat{n}\perp q$ still permits $\hat{v}\perp\hat{n}$ for arbitrary angle between H and q, with a residual degeneracy that might serve, for example, to accommodate the presence of walls. In the axial state, the vector $\hat{\ell}$ lies along q, and F_D is lowest when $\hat{v}$ lies perpendicular to H in the plane containing H and q. A more general procedure for the axial state treats F_K, F_Z, and F_D together; as expected, the configuration deforms continuously to $\hat{v}\perp H$, $\hat{\ell}\,||\,v$ and $\hat{\ell}\,||\,q$, $\hat{v}\,||\,\hat{\ell}$ as $q \to 0$ or $H \to 0$, respectively. In the former case ($Kq^2 \lesssim g_D$), the squared gap has its symmetry axis perpendicular to H owing to the dipole coupling between $\hat{v}$ and $\hat{\ell}$. For simplicity, we omit the similar but more complicated analysis for the two-dimensional state.

The remaining state of experimental interest is the three-dimensional one. For H = 0 and q = 0, the two orthonormal triplets ($\hat{v}$'s and $\hat{n}$'s) yield an isotropic squared energy gap, with no orientational freedom to eliminate the magnetic or kinetic energy. Hence, application of these external perturbations deforms the state markedly. If the dipole energy is omitted, the nonlinear equation for $A_{\mu i}$ in (1) is easy to derive but difficult to solve, owing to the different action of H and q on the indices of $A_{\mu i}$.

To construct the general solution, it is helpful first to consider the limits q = 0 or H = 0. In the former case, we choose $\hat{z}$ along H and take $A_{\mu i}$ in the form $(\Lambda^{1/2}R)_{\mu i}$, where R is an arbitrary rotation matrix, and Λ is diagonal with elements $\lambda_1, \lambda_1, \lambda_3$ ($\lambda_3 < \lambda_1$). In this way, the squared energy gap $\hat{k}_i(R\Lambda R)_{ij}\hat{k}_j$ acquires a flattened shape with the symmetry axis in the direction RH.[15] The second simple limiting case occurs for H = 0, when $\hat{z}$ is chosen to lie along q. The order parameter $A_{\mu i}$ has the form $(R\Lambda^{1/2})_{\mu i}$ with R again arbitrary; this choice orients the flattened symmetry axis of $|\Delta(\hat{k})|^2$ along q, independent of R. For both these special cases, adding the dipole energy as a weak perturbation to H or to q fixes the axis of R to lie along $\hat{H}$ or $\hat{q}$, and F_D is minimized for a rotation angle $\arccos[-\frac{1}{4}(\lambda_3/\lambda_1)^{1/2}]$. In the limit of weak perturbations ($g_Z H^2 + Kq^2 << \alpha$), we show below that $\lambda_3 \to \lambda_1$, thereby recovering Leggett's value $\arccos(-\frac{1}{4})$.[11]

These examples are relatively simple, because the choice q = 0 or H = 0 leaves only a single preferred direction. When both perturbations dominate the dipole interaction, however, the problem becomes more intricate. It is convenient to imagine an arbitrary fixed coordinate system, which serves as a reference frame for both $\underset{\sim}{H}$ and $\underset{\sim}{q}$. Moreover, let R_H and R_q denote the matrices that rotate $\underset{\sim}{H}$ and $\underset{\sim}{q}$ to lie along the $\hat{z}$ axis [i.e., $(R_H)_{ij}H_j = H\delta_{iz}$]. Then it is not difficult to verify that the assumption

$$A_{\mu i} = (\tilde{R}_H R \Lambda^{1/2} R_q)_{\mu i} = (\tilde{R}_H \Lambda^{1/2} R\, R_q)_{\mu i} \tag{15}$$

satisfies the nonlinear field equation identically. Here R is a rotation about the $\hat{z}$ axis, and Λ is again diagonal with elements λ_1, λ_1, λ_3; evidently, Λ is invariant under rotations about $\hat{z}$ and therefore commutes with R. A detailed calculation yields the equilibrium parameters

$$\lambda_1 = \frac{1}{2}(3\beta_{12} + \beta_{345})^{-1}[\alpha + \beta_{345}{}^{-1}(2\beta_{12} - \beta_{345})Kq^2 + \beta_{345}{}^{-1}\beta_{12}g_Z H^2] \tag{16}$$

$$\lambda_3 = \frac{1}{2}(3\beta_{12} + \beta_{345})^{-1}[\alpha - \beta_{345}{}^{-1}(4\beta_{12} + 3\beta_{345})Kq^2 - \beta_{345}{}^{-1}(2\beta_{12} + \beta_{345})g_Z H^2]$$

and the corresponding free energy

$$F = -\frac{(\alpha - \frac{5}{3}Kq^2 - \frac{1}{3}g_Z H^2)^2}{4(\beta_{12} + \frac{1}{3}\beta_{345})} - \frac{(2Kq^2 + g_Z H^2)^2}{6\beta_{345}} \; . \tag{17}$$

These expressions generalize those for H = 0 or q = 0; they demonstrate explicitly the hydromagnetic deformation of the three-dimensional state. For example, application of the thermodynamic identity $\chi = -(\partial^2 F/\partial H^2)|_{H=0}$ to Eq. (17) implies that the static susceptibility χ_{3d} is less than χ_n; moreover, χ_{3d} at fixed flow velocity depends explicitly on v or q. A similar situation would arise at fixed current J. The squared energy gap $|\Delta(\hat{k})|^2 = \hat{k}_i(\tilde{R}_q \Lambda R_q)_{ij}\hat{k}_j$ is axially symmetric about $\hat{q}$ and independent of $\hat{H}$, but its values λ_1 and λ_3 perpendicular and parallel to $\hat{q}$ depend on H^2 as well as q^2.

Inclusion of the dipole energy as a weak perturbation to the magnetic and kinetic energy partially lifts the residual degeneracy of the three-dimensional state. If the axis of R lies in the plane of $\underset{\sim}{q}$ and $\underset{\sim}{H}$, then F_D is minimized for a rotation angle

$$\arccos\left[\left(\frac{\lambda_3}{\lambda_1}\right)^{1/2} \frac{1-2\cos\psi}{2(1+\cos\psi)}\right] \tag{18}$$

if the quantity in brackets is less than 1, and zero otherwise, where ψ is the smaller angle between $\hat{H}$ and $\pm \hat{q}$. Note that ψ affects the minima of F_D significantly; detailed analysis shows similar changes in the functional dependence on the rotation angle, both of which persist even if $\lambda_3 \approx \lambda_1$. Such alterations in F_D should lead to unusual NMR signals, which deserve additional study.

The mass current in the three-dimensional state becomes

$$\underset{\sim}{J} = \frac{10K\underset{\sim}{q}}{h(3\beta_{12}+\beta_{345})} \left[\alpha - \left(\frac{8\beta_{12}+11\beta_{345}}{5\beta_{345}}\right) Kq^2 - \left(\frac{4\beta_{12}+3\beta_{345}}{5\beta_{345}}\right) g_Z H^2\right] \tag{19}$$

which follows either from Eq. (4) or from the relation $\delta F = m\underset{\sim}{J} \cdot \delta\underset{\sim}{v} = \frac{1}{2} h\underset{\sim}{J} \cdot \delta\underset{\sim}{q}$. It depends explicitly on the magnitude (but not the direction of H [compare Eq. (14)]. With increasing $\underset{\sim}{q}$, Eq. (19) reaches a maximum value J_c at a critical value q_c, determined by the relation

$$5\alpha\beta_{345} = 3(8\beta_{12} + 11\beta_{345})Kq_c^2 + (4\beta_{12} + 3\beta_{345})g_Z H^2 \ . \tag{20}$$

Thus q_c and J_c now depend on H as well as T. At fixed H, the critical current vanishes above a temperature (we here use the weak-coupling value for α)

$$T^*(H) = T_c \{1 - \tfrac{3}{5}[g_Z H^2/N(0)][4(\beta_{12}/\beta_{345}) + 3]\} \ , \tag{21}$$

and J_c varies like $[T^*(H) - T]^{3/2}$ for $T < T^*(H)$. Observation of this magnetic suppression of hydrodynamic flow would help confirm the identification of the B phase as three-dimensional.

The hydromagnetic deformation of the three-dimensional state implies a corresponding shrinkage of the ^{3}He-B portion of the experimental phase diagram. Assume $\beta_{345} > \frac{3}{2}\beta_{13}$ to ensure that the three-dimensional state is stable at $q = 0$ and $H = 0$. Within this limited domain (still taking β_{13} and β_{345} positive), there are several different possibilities.

(1) $\beta_{345} > 2\beta_{13}$, which makes the two-dimensional state stable relative to the axial one. The three-dimensional state then undergoes a phase transition for nonzero q and H to a two-dimensional one, with the precise mechanism depending on the ratio $g_Z H^2/Kq^2$. Either the mass current exceeds its critical value, giving a first-order transition along line 1 in Fig. 1a, or the three-dimensional state deforms continuously to the two-dimensional one when $\lambda_3 = 0$, giving a second-order transition along line 2 in Fig. 1a. Note that the order of the transition depends on the ratio $g_Z H^2/Kq^2$. Unfortunately, neither of these transitions is likely to occur in

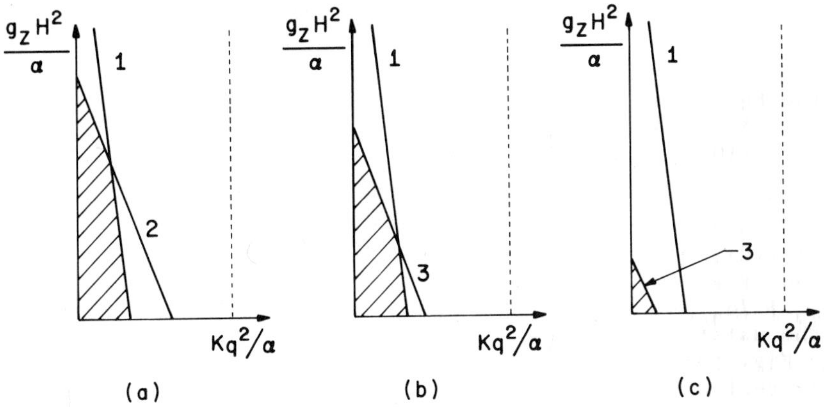

Figure 1. Hydromagnetic phase diagram of ^{3}He, with the three-dimensional state stable in the shaded region. The onset of critical current occurs along line 1 for the three-dimensional state [Eq. (20)] and along the dotted line for the two-dimensional and axial states, beyond which the system is normal.

(a) $\beta_{345} > 2\beta_{13}$: The equivalent conditions $\lambda_3 = 0$ [see Eq. (16)] and $F_{3d} = F_{2d}$ occur along line 2, which represents a second-order transition to the two-dimensional state.

(b) and (c) $2\beta_{13} > \beta_{345} > \frac{3}{2}\beta_{13}$: The condition $F_{3d} = F_{ax}$ occurs along line 3, which represents a first-order transition to the axial state. Lines 3 and 1 intersect if $\frac{1}{2}(2\beta_{12} + 9\beta_{345})(6\beta_{12} + 7\beta_{345})^{-1} > f(\beta) > 0$ (shown in b), whereas they do not intersect if $1 > f(\beta) > \frac{1}{2}(2\beta_{12} + 9\beta_{345})(6\beta_{12} + 7\beta_{345})^{-1}$ (shown in c), where $f(\beta)$ is defined in Eq. (23).

practice because the weak-coupling limit already lies at the boundary of the allowed domain [see Eq. (3)], and spin fluctuations appear to diminish the ratio β_{345}/β_{13} toward and beyond the value 3/2.[4] Thus we turn to the second alternative:

(2) $2\beta_{13} > \beta_{345} > \frac{3}{2}\beta_{13}$, where the competition is between the three-dimensional state (at H = 0 and q = 0) and the axial state. The associated transition is always first order, occurring either because J exceeds its critical value or because the free energies of the two states become equal. The first possibility is shown as line 1 in Figs. 1b and 1c; the second possibility $F_{3d} = F_{ax}$ defines a line

$$[1-f(\beta)]\alpha = [4(\beta_{12}/\beta_{345})+3-f(\beta)]kq^2 + [2(\beta_{12}/\beta_{345})+1]g_z H^2 \quad (22)$$

in the $Kq^2 - g_Z H^2$ plane (labeled 3 in Figs. 1b and 1c), where

$$f(\beta) = [(3\beta_{12} + \beta_{345})(2\beta_{13} - \beta_{345})/\beta_{345}\beta_{245}]^{1/2} \tag{23}$$

varies continuously from 0 at $\beta_{345} = 2\beta_{13}$ to 1 at $\beta_{345} = \frac{3}{2}\beta_{13}$. For small $f(\beta)$, the critical current can actually be reached at small values of $g_Z H^2/Kq^2$ (see Fig. 1b), but for $2f(\beta) > (2\beta_{12} + 9\beta_{345})\cdot(6\beta_{12} + 7\beta_{345})^{-1}$, the transition from the three-dimensional state to the axial one always takes place because of equal free energy (see Fig. 1c). This last situation also occurs throughout the whole region $2\beta_{13} > \beta_{345} > \frac{3}{2}\beta_{13}$ if $g_Z H^2/Kq^2$ is sufficiently large, which describes all previous experiments.

The preceding discussion provides a basis for understanding the observed appearance of the A phase at all pressure in finite magnetic field.[1,16] With the usual identification of the B and A phases as three-dimensional and axial, Eq. (22) predicts that the BA transition should occur at a field-dependent temperature T_{AB} defined by the relation

$$[1 - f(\beta)]\,\frac{T_c - T_{AB}}{T_c} = \frac{7\zeta(3)}{16\pi^2}\left(\frac{\gamma\hbar H}{k_B T_c}\right)^2 \frac{2\beta_{12} + \beta_{345}}{\beta_{345}} \tag{24}$$

where we have used the weak-coupling values of α and g_Z, and taken $q = 0$ for simplicity. Well below the polycritical pressure, the quantity $1 - f(\beta)$ is nonzero, and the width of the A-phase strip varies quadratically with H. Moreover, the latent heat at the BA transition should become small at low pressure, since the strict weak-coupling limit predicts a second-order transition, discussed above. As p approaches the polycritical value along the line T_c from below, however, $1 - f(\beta)$ vanishes linearly and must itself be expanded to first order in $T_c - T_{AB}$. Hence the width of the A strip should vary linearly with H near the polycritical point. Above this pressure, the B phase is stable only for temperatures below the zero-field line T_{AB}, which must be identified as the locus of points where $1 = f(\beta)$. Thus Eq. (24) takes the slightly different form above the polycritical point

$$\left(\frac{df}{dT}\right)_{T^0_{AB}} (T^0_{AB} - T_{AB})\left[\frac{T_c - T^0_{AB}}{T_c} + \frac{T^0_{AB} - T_{AB}}{T_c}\right] = \frac{7\zeta(3)}{16\pi^2}\left(\frac{\gamma\hbar H}{k_B T_c}\right)^2 \frac{2\beta_{12} + \beta_{345}}{\beta_{345}} \tag{25}$$

and the H-dependent shift in the AB boundary should change continuously from linear to quadratic with increasing $T_c - T^0_{AB}$. This description agrees qualitatively with the approximately hyperbolic experimental shape.[1,16] Finite flow in small or zero magnetic fields should produce similar effects, whose magnitude can, in principle, be fixed by the available data.[1,16] One new feature at low fields is that J should actually reach its critical value for sufficiently low pressure, initiating the BA transition even though $F_{3d} < F_{ax}$; such a change in mechanism might be detectable as an altered latent heat.

We have examined the hydromagnetic properties of superfluid ^{3}He. The predicted effects differ qualitatively from those in the normal phase, and indeed from those in other materials. Experimental study of these phenomena would be most desirable.

ACKNOWLEDGMENTS

I am grateful to D. A. Dahl and J. W. Serene for many valuable discussions.

References

(1) A complete bibliography would be prohibitive, and we instead refer to J. C. Wheatley, Rev. Mod. Phys. 47, 415 (1975).

(2) R. Balian and N. R. Werthamer, Phys. Rev. 131, 1553 (1963).

(3) N. D. Mermin and G. Stare, Phys. Rev. Lett. 30, 1135 (1973).

(4) W. F. Brinkman and P. W. Anderson, Phys. Rev. A8, 2732 (1973); W. F. Brinkman, J. W. Serene, and P. W. Anderson, Phys. Rev. A10, 2386 (1974).

(5) V. Ambegaokar, 1974 Canadian Association of Physicists, International Summer School in Theoretical Physics (unpublished).

(6) P. G. deGennes, Phys. Lett. 44A, 271 (1973); V. Ambegaokar, P. G. deGennes, and D. Rainer, Phys. Rev. A9, 2676 (1974); P. G. deGennes and D. Rainer, Phys. Lett. 46A, 429 (1974).

(7) E. I. Blount (unpublished).

(8) D. Dahl (unpublished).

(9) L. D. Landau and E. M. Lifshitz, Electrodynamics of Continuous Media (Addison-Wesley, Reading, Mass., 1960), Secs. 30 and 31.

(10) V. Ambegaokar and N. D. Mermin, Phys. Rev. Lett. 30, 81 (1973).

(11) A. J. Leggett, Ann. Phys. (N.Y.) 85, 11 (1974).

(12) C. M. Varma and N. R. Werthamer, Phys. Rev. A9, 1465 (1974).

(13) N. D. Mermin and V. Ambegaokar, in Collective Properties of Physical Systems, edited by B. Lundquist and S. Lundquist (Academic Press, N.Y., 1974), pp. 97-102.

(14) P. G. deGennes, Superconductivity of Metals and Alloys (Benjamin, N.Y., 1966), pp. 182-185.

(15) S. Engelsberg, W. F. Brinkman, and P. W. Anderson, Phys. Rev. A9, 2592 (1974).

(16) D. N. Paulson, H. Kojima, and J. C. Wheatley, Phys. Rev. Lett. 32, 1098 (1974).

SUPERFLUID DENSITY OF ^{3}He *

H. Kojima, D. N. Paulson, and J. C. Wheatley

Department of Physics,
University of California at San Diego
La Jolla, California 92037

Abstract: Measurements of superfluid density in ^{3}He via fourth sound in an open parallel plate "superleak" show an unexpected pressure and pore size dependence and an unexpectedly small change at the B → A transition.

We present the first measurements of fourth sound in ^{3}He using a resonator with a parallel plate geometry. The fourth sound velocity c_4 is used to obtain a relative superfluid density by the equation $\rho_s/\rho = c_4{}^2/c_1{}^2$, where c_1 is the first sound velocity measured in the same resonator.[1,2] The measurements (H_o < 0.1 gauss) cover a pressure range from 11 to 33 bar in both A and B phases and show a change at the B → A transition not evident in earlier work[3,4] with a confined superleak. This change reflects on the possible orbital anisotropy of the fluids. We find a surprising pressure dependence of the effect of superleak pore size on superfluid density and can comment quantitatively on strong coupling effects and the efficacy of the "molecular field" concept[5] in dealing with some effects of the Fermi liquid, finding a surprising dependence of "strong coupling" parameters on pressure.

The epoxy resonator consisted of a number of parallel channels each 50 microns deep, 5 mm wide, and 20 mm long filled with ^{3}He cooled by adiabatic demagnetization of CMN. Provisional absolute temperatures T were obtained from magnetic temperatures measured outside the resonator via thermal observations of the second order

*Supported by the U. S. Atomic Energy Commission under Contract number AT(04-3)-34, P.A. 143.

transition at a variety of pressures.[6] Near T_c we found ρ_s/ρ to be linear in T to the experimental accuracy, so to eliminate approximately the small temperature differences due to heat flow between the resonator and the CMN thermometer and coolant we extrapolated ρ_s/ρ linearly to zero to obtain the T_c used in the reduced temperature plots. We observed both the first longitudinal resonance f_1 and a lower frequency f_L. The ratio f_1/f_L was measured to be independent of temperature and pressure and did not change at the $B \to A$ transition. Since f_L could be measured more precisely and closer to T_c than f_1, most of our measurements refer to it, but none of our qualitative conclusions is altered if only f_1 data are used. The effect of noninfinite viscous penetration depth is calculated to have an insignificant effect on the results. Details of the resonator and of the experiments will be presented elsewhere.

Data for both the "open" (50 micron) and "confined" (pore size probably of order one micron) geometries of the parallel plate and CMN superleak[3] are shown in Figure 1. Opening up the superleak makes ρ_s/ρ linearly dependent on T near T_c and increases ρ_s/ρ. These effects are small near the melting pressure and larger at the lowest pressure. There is very little effect of pressure on ρ_s/ρ for the confined geometry, as observed also by Yanof and Reppy.[4] Thus ρ_s/ρ at any pressure for the confined geometry is about the same as that of ^{3}He-A in the open geometry near melting pressure. A $B \to A$ transition for 27.6 bar is shown in Figure 1. The transition is broad but occurs near the expected temperature. Further details on pressure dependence are given in Figure 2, where from top to bottom we show the initial slope, the fractional density change at $B \to A$, and the reduced temperature difference of the $B \to A$ transition as taken from the centers of the present transition and from other data.[7-10] The fractional density change at $B \to A$ is imprecise owing to the broad transition and the small change, but it appears to have no more than a weak pressure dependence. We find $\rho_{s_A} > \rho_{s_B}$ while Alvesalo et al.[11] found $\rho_{s_A} < \rho_{s_B}$ in their melting pressure vibrating wire experiments at 1490 gauss. If one or both of the fluids is anisotropic then either sign for the density change is possible. Considering the broad $B \to A$ transition, its center temperature agrees favorably with the earlier data.

The fractional change of (ρ_s/ρ) at the $B \to A$ transition has the expected sign but is unexpectedly small. Near the PCP and near T_c we expect the averaged ρ_s/ρ to be the same for both A and B phases. But if B is an orbitally isotropic fluid and A has the angular properties of the ABM state[12] with the orbital orientation vector $\vec{\ell}$ perpendicular to the plates[13] then we would expect[14,15] ρ_s/ρ to increase by a factor 6/5 on $B \to A$. The vector $\vec{\ell}$ can be disoriented by superflow parallel to the plates either from the sound field itself or from hydrodynamic heat flow. There is no effect on ρ_s/ρ of sound amplitude, excepting heating effects. The maximum superfluid

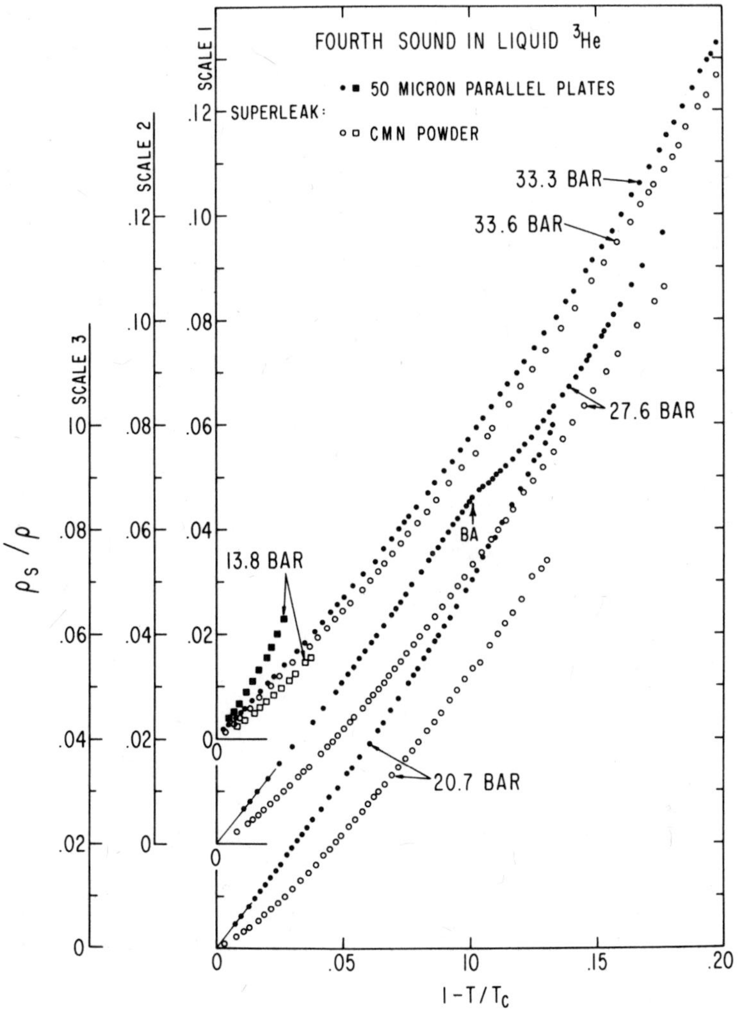

Figure 1. Superfluid density as deduced from fourth sound measurements with a 50 micron parallel plate (closed symbols) and a CMN powder (open symbols) superleak at several pressures as a function of reduced temperature difference. (↑BA) is the location of the expected B → A transition at 27.6 bar.

velocity expected from heat flow is less than a few percent of $h/2m_3d$, where d is the plate separation; so no disorientation is expected.[16] Another possible interpretation of the small density change is that the B phase is not orbitally isotropic, so that only a change of

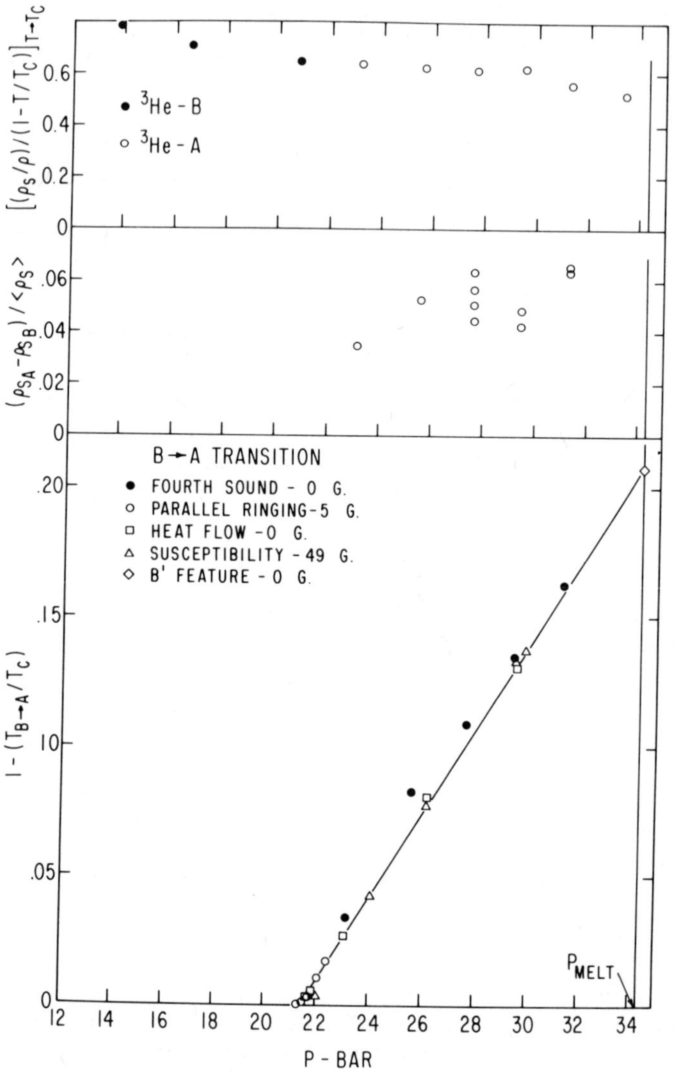

Figure 2. Pressure dependence of the initial slope of (ρ_s/ρ) vs $(1-T/T_c)$, the fractional superfluid density change at the B → A transition, and the reduced temperature difference of the B → A transition (fourth sound, present data; parallel ringing, Ref. 9; heat flow, Ref. 7; susceptibility, Ref. 8; B' feature, Ref. 10).

anisotropy is observed. The vibrating wire experiments of Alvesalo et al.[11] in the B phase near $T_{B\to A}$ when interpreted in terms of a scalar ρ_n/ρ and normal viscosity give values of $(\rho_s/\rho)_B$ very much larger than might be extrapolated from the present higher pressure measurements. Since their experiments are accurate enough we feel that the discrepancy calls into question the assumption of orbital isotropy for the B phase, particularly in the sense that it should not be assumed without proof.

Very near T_c the superfluid density is thought to depend on temperature and pressure by the formula[6] $\rho_s/\rho = sA(1-T/T_c)/(1+F_1/3)$, where A is a parameter that can be calculated from the assumed microscopic state of the ^{3}He, s is a strong coupling parameter given by $s = (C_</C_>-1)_{expt.}/(C_</C_>-1)_{weak}$, and $(1+F_1/3)$ is a "molecular field" correction.[5] Here A is 2 both for the isotropic BW state[17] and the perpendicular component of the ABM state,[15] $C_</C_>$ is the ratio of specific heat just below to just above T_c, and $(C_</C_> - 1)_{weak}$ is 1.43 for the BW state and 1.43/(6/5) for the ABM state. In Figure 3 we plot $sA = [(\rho_s/\rho)/(1-T/T_c)](1+F_1/3)$ both from the initial slopes in Figure 2 for the open geometry and from the value of ρ_s/ρ at $(1-T/T_c) = .05$ obtained[3] in the confined (CMN) geometry. For the latter only the general variation with pressure should be taken seriously. Also shown are values of 2s for the ABM state using the empirical heat capacity measurements of Webb et al.[18] obtained for ^{3}He in the same kind of confined geometry as that of the CMN ρ_s/ρ measurements. Since the results are very dependent on the local accuracy of the temperature scale we show also on Figure 3 some unpublished parallel ringing frequencies of Webb, Kleinberg, and Wheatley which used the same temperature scale to find quite a reasonable pressure dependence.[6] Near the PCP the strong coupling factor for the B phase would be about one on the basis of the specific heat measurements. From Figure 3 we deduce that near the melting pressure, where there is little effect of confinement on ρ_s/ρ, the observed value of (sA) from the superfluid density is in reasonable agreement with the measured "confined" specific heat and the assumption of the oriented ABM state. But as the pressure drops, (sA) is much larger than expected for the open geometry.

The above observations motivate several questions: 1) Is $C_</C_>$ the same in bulk ^{3}He as in the pores of CMN or is it larger at lower pressures as may be suggested by the data of Figure 3? From ultrasonic measurements[19] T_c appears to be unshifted by confinement to a precision of order 0.1%. 2) Does a BA transition occur in the confined geometry even though it is not observable via ρ_s/ρ? A transition is observed in 12 micron pores by Ahonen et al.[20,21] 3) Does ^{3}He in a confined geometry at a given T and P have the same properties as the corresponding "bulk" fluid, aside from effects on a length scale of order $\xi_0(1-T/T_c)^{-1/2}$? The super-

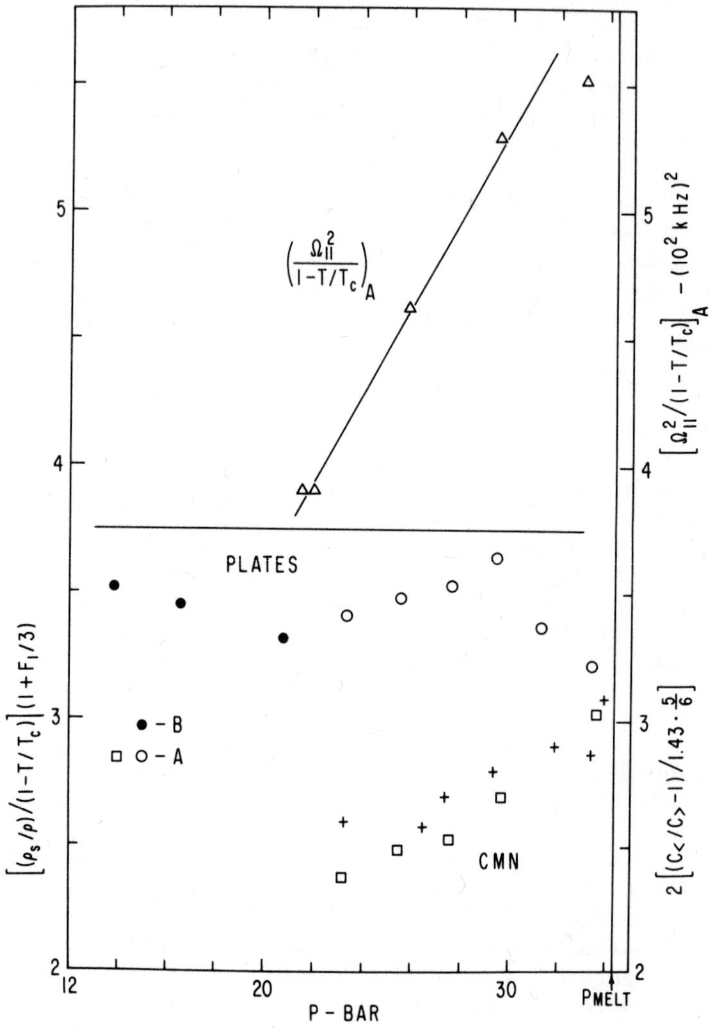

Figure 3. Plot showing the effect of pressure and confinement on "strong coupling" effects in ρ_s/ρ. (O, ^{3}He-A and ●, ^{3}He-B in plate superleak; □, ^{3}He-A in CMN superleak; + , values of sA (see text) from confined specific heat measurements). The ringing frequency data are shown to provide a confidence level for the temperature scale.

fluid density measurements near melting pressure would answer yes, but the present superfluid density measurements and perhaps the dynamic magnetism measurements[20,21] suggest that the answer is no at lower pressures.

We have also studied, for the B phase, the effects of the molecular fields and the strong coupling factor s in correlating and predicting superfluid density and susceptibility measurements using

$$\frac{\rho_s}{\rho} = \frac{(1 - Y)/(1 + F_1/3)}{1 - \frac{F_1/3}{1 + F_1/3}(1 - Y)} \tag{1}$$

and

$$\frac{\chi_N - \chi_B}{\chi_N} = \frac{\frac{1}{3}\frac{1 + F_1/3}{1 + Z_0/4}}{1 + \left(\frac{F_1}{3} + \left|\frac{Z_0}{4}\right| \frac{1}{3}\frac{1 + F_1/3}{1 + Z_0/4}\right)\frac{\rho_s}{\rho}} \frac{\rho_s}{\rho}, \tag{2}$$

where F_1 and Z_0 are Landau parameters,[6] and the B phase is assumed to have 1/3 nonmagnetic pairs. Our B phase data as a function of T/T_c at 20.7 bar are satisfactorily predicted by (1) if, following Combescot's suggestion,[22] Y is the Yosida function for T_e/T_c; where T_e/T_c is found by solving $s^{1/2}\,\Delta_{BCS}\,(T/T_c)/(T/T_c) = \Delta_{BCS}(T_e/T_c)/(T_e/T_c)$ with s adjusted to fit the ρ_s/ρ data near T_c. Also, if we assume that ρ_s/ρ in the B phase at 33.3 bar is 6% less (Figure 2) than that measured for the A phase, then the susceptibility predicted for a metastable B phase at 33.3 bar by Eq. (2) extrapolates to a value only a few percent less than the bulk measurements at melting pressure of Corruccini and Osheroff[23] who made a similar comparison based on our high pressure A phase ρ_s/ρ data in a confined geometry[3] and reached a similar conclusion. However, our 20.8 bar susceptibility data[24] are not predicted by the 20.7 bar ρ_s/ρ data and Eq. (2),[6,23] the predicted $(\chi_N-\chi_B)/\chi_N$ being about 3/4 of that measured and apparently well outside the possible error in the calibration constant, which is interlocked with the diamagnetic susceptibility.

We thank Dr. D. Osheroff for bringing Combescot's work to our attention and Prof. I. Rudnick for useful discussions.

References

(1) K. A. Shapiro and I. Rudnick, Phys. Rev. A 137, 1383 (1965).

(2) R. Graham, Phys. Rev. Lett. 33, 1431 (1974).

(3) H. Kojima, D. N. Paulson, and J. C. Wheatley, Phys. Rev. Lett. 32, 141 (1974).

(4) A. W. Yanof and J. D. Reppy, Phys. Rev. Lett. 33, 631 (1974).

(5) A. J. Leggett, Phys. Rev. A 140, 1869 (1965).

(6) J. C. Wheatley, Rev. Mod. Phys. 47, 415 (1975).

(7) T. J. Greytak, R. T. Johnson, D. N. Paulson, and J. C. Wheatley, Phys. Rev. Lett. 31, 452 (1973).

(8) D. N. Paulson, R. T. Johnson, and J. C. Wheatley, Phys. Rev. Lett. 31, 746 (1973).

(9) R. L. Kleinberg, D. N. Paulson, R. A. Webb, and J. C. Wheatley, J. Low Temp. Phys. 17, 521 (1974).

(10) D. D. Osheroff, private communication.

(11) T. A. Alvesalo, H. K. Collan, M. T. Loponen, and M. C. Veuro, Phys. Rev. Lett. 32, 981 (1974).

(12) P. W. Anderson and W. F. Brinkman, Phys. Rev. Lett. 30, 1108 (1973).

(13) V. Ambegaokar, P. G. de Gennes, and D. Rainer, Phys. Rev. A 9, 2676 (1974).

(14) W. M. Saslow, Phys. Rev. Lett. 31, 870 (1973).

(15) R. Combescot, preprint.

(16) P. G. de Gennes and D. Rainer, Phys. Lett. 46A, 429 (1974).

(17) R. Balian and N. R. Werthamer, Phys. Rev. 131, 1553 (1963).

(18) R. A. Webb, T. J. Greytak, R. T. Johnson, and J. C. Wheatley, Phys. Rev. Lett. 30, 210 (1973).

(19) D. N. Paulson, R. T. Johnson, and J. C. Wheatley, Phys. Rev. Lett. 30, 829 (1973).

(20) A. I. Ahonen, M. T. Haikala, and M. Krusius, Phys. Lett. 47A, 215 (1974).

(21) A. I. Ahonen, M. T. Haikala, M. Krusius, and O. V. Lounasmaa, Phys. Rev. Lett. (in press).

(22) R. Combescot, preprint.

(23) L. R. Corruccini and D. D. Osheroff, preprint.

(24) D. N. Paulson, H. Kojima, and J. C. Wheatley, Phys. Lett. 47A, 457 (1974).

COMMENTS ON THE JOSEPHSON PLASMA RESONANCE IN $^3He(A)$*

A. J. Dahm† and D. N. Langenberg§

†Department of Physics, Case Western Reserve University
Cleveland, Ohio 44106

§Department of Physics and Laboratory for Research
on the Structure of Matter, University of Pennsylvania
Philadelphia, Pennsylvania 19174

Abstract: Phenomena relating to the longitudinal nuclear magnetic resonance in superfluid ^3He A are discussed on the basis of a previously proposed analogy between this resonance and the Josephson plasma resonance in superconductors. Some effects which are subject to experimental test are suggested.

The interesting phenomenon of a longitudinal nuclear magnetic resonance[1-3] in the A phase of superfluid ^3He was predicted by Leggett,[4] who described this resonance as an internal Josephson effect[5] between the two condensates containing respectively the S_Z equal +1 and -1 components of helium atoms paired in the triplet state. The z axis is taken to be parallel to an applied magnetic field. Maki and Tsuneto[6] reformulated the Josephson equations describing this resonance in terms of parameters analogous to those used to describe a superconducting Josephson junction. This made the analog between the longitudinal nuclear magnetic resonance and the Josephson plasma mode much more explicit. We call attention here to effects associated with the Josephson plasma resonance in tunnel junctions and suggest how these effects might be observed in the case of $^3He(A)$.

In order to suggest these analogs it is necessary to write the

*Work partially supported by the Army Research Office (Durham) and the National Science Foundation.

Josephson equations for the two systems. For the superconducting case[7] these are

$$j = j_1 \sin\phi + GV\,(1 + \zeta\cos\phi) \tag{1}$$

$$\hbar\, d\phi/dt = -\,2\Delta\mu \;=\; 2eV \tag{2}$$

Here j is the tunneling current; j_1 is the Josephson critical current; ϕ is the difference in the phases of the order parameters describing the two pair condensates (superconductors); G is a conductance constant; V is the voltage across the junction; ζ is a constant approximately equal to -0.9; and $\Delta\mu$ is the difference in the chemical potential for electrons in the two condensates. The contribution to the current $GV\zeta\cos\phi$ is the quasiparticle-pair interference current.[8-10]

The analogous equations describing ^{3}He(A) are[6]

$$dn/dt = j_0 \sin\phi - (n - n_0)/T_1 \tag{3}$$

$$\hbar\, d\phi/dt = -\,2(\mu_\uparrow - \mu_\downarrow) + 8\,I\,n \tag{4}$$

Here $4n = n_\uparrow - n_\downarrow$, where $n_\uparrow$ and $n_\downarrow$ are the numbers of spin-up and spin-down atoms; j_0 is proportional to the square of the order parameter and depends on the strength of the dipole coupling; ϕ is the difference in the phases of the order parameters describing the two condensates; n_0 is the equilibrium value of n in a static applied field; T_1 is the spin relaxation time; $\mu_\uparrow$ and $\mu_\downarrow$ are the chemical potentials of the spin-up and spin-down atoms in the absence of the exchange interaction; I is the spin exchange coupling constant and is negative since the coupling is antiferromagnetic. In the presence of an applied magnetic field H, $\mu_\uparrow - \mu_\downarrow = -\hbar\gamma H$, where γ is the gyromagnetic ratio of a ^{3}He atom ($\gamma < 0$). The value of n_0 is given by a steady state solution to Eq. (4). The current dn/dt is the rate of pair transfer from one condensate to another as in the superconducting case.

We next examine the small amplitude plasma resonance frequency for the two cases. We will consider only pair currents for expediency. Consider a superconducting Josephson junction with a capacitance C carrying a dc current $j_1 \sin\phi_0$ in the zero voltage mode ($d\phi/dt = 0$). An excess charge q placed on one side of the junction results in a voltage $V = q/C$ which causes the phase to change. This phase change in turn results in an additional current via Eq. (1) which discharges the capacitor. The pair current is

$$j = j_{dc} - \dot{q} \quad , \tag{5}$$

or rewriting

$$\dot{q} = - j_1 \sin \phi + j_1 \sin \phi_o \quad . \tag{6}$$

In the absence of losses, the entire excess charge is transferred across the oxide barrier before the capacitive component of the current reverses sign. The plasma resonance consists of an oscillatory pair current across the junction with the phase executing small oscillations about a steady state value,

$$\phi = \phi_o + \delta\phi(t) \tag{7}$$

The equation of motion for ϕ, obtained by differentiating Eq. (2) and using Eq. (6) is

$$\ddot{\phi} + \omega_J^2 \sin \phi - \omega_J^2 \sin \phi_o = 0 \quad , \tag{8}$$

where $\omega_J^2 = 2ej_1/\hbar C$ is the Josephson plasma frequency. The small amplitude solution to Eq. (8) is a harmonic oscillation for $\delta\phi$ with angular frequency $\omega_P = \omega_J \cos^{1/2}\phi$. Equation (8) has been discussed in the literature as an analog of the torqued pendulum.[11-13]

The derivation of an equation identical to Eq. (8) for the longitudinal magnetic resonance (plasma resonance) in superfluid ^{3}He(A) in the presence of a dc "bias current," $j_o \sin \phi_o$, follows by analogy with $\omega_J^2 = |8\ I\ j_o|/\hbar$. The excess number of spins 4n is the analog of the charge q, and the antiferromagnetic coupling provides the restoring force to drive n towards its equilibrium value n_o. Maki and Tsuneto[6] have discussed Eq. (8) in the limit $\phi_o = 0$.

We make the following comments based on this analogy.

1) The longitudinal magnetic resonance frequency will vary as $\cos^{1/2}\phi_o = [1 - (j_o^{-1}\ dn/dt)^2]^{1/4}$. This dependence on dc bias current has been verified experimentally in superconducting tunnel junctions.[14]

2) We suggest a relatively simple method for generating a dc bias current. In the presence of an applied magnetic field which increases linearly in time, $H = \beta t$, there exists a steady state solution to Eq. (4) in which dn/dt is constant. The damping term in Eq. (3) allows the system to arrive at a steady state current given by $dn/dt = - \hbar\gamma\beta/4I = j_o \sin\ \phi_o$. We rewrite this equation as $\sin \phi_o = - \hbar\gamma\beta/4Ij_o = 2\gamma\beta/\omega_J^2$ and observe that to obtain a value of $\sin \phi_o$ of order unity, requires $\beta \simeq 10^5 - 10^6$ G/sec, an easily attainable rate of change of field. A measurement of the longitudinal magnetic resonance with its predicted dependence on dc current could be used to detect such a current. Note that while a large value of β is required, the resonance frequency can be measured in a msec using the ringing technique of Webb _et al._[3]

3) In the theory of a Josephson tunnel junction, the sinϕ pair current and the cos ϕ quasiparticle-pair-interference current arise as Kramers-Kronig-conjugate real and imaginary parts of a single response function.[15] One cannot have one without the other. It would be interesting to determine experimentally whether a phase-dependent damping term[16] in addition to the phenomenological relaxation term of Eq. (3) exists, that is, whether ζ or its equivalent is non-zero. In the superconducting case, the value of ζ is determined by measuring the resonance Q as a function of dc bias current. It would be much simpler in ^{3}He(A) to compare the damping of the large amplitude oscillations over the first few cycles (large excursions of cos ϕ) with the small amplitude damping in a ringing experiment.

4) An analysis of the plasma resonance in the presence of a large rf field[17] predicts a branching of the fundamental resonance into two resonances and the generation of fractional as well as whole harmonics. A plot of some of these resonances in the dc current-rf voltage amplitude plane is shown in Fig. 1. These effects have been observed in superconducting tunnel junctions.[14,17-19] Similar effects in ^{3}He(A) should be observable in the presence of large rf magnetic fields.

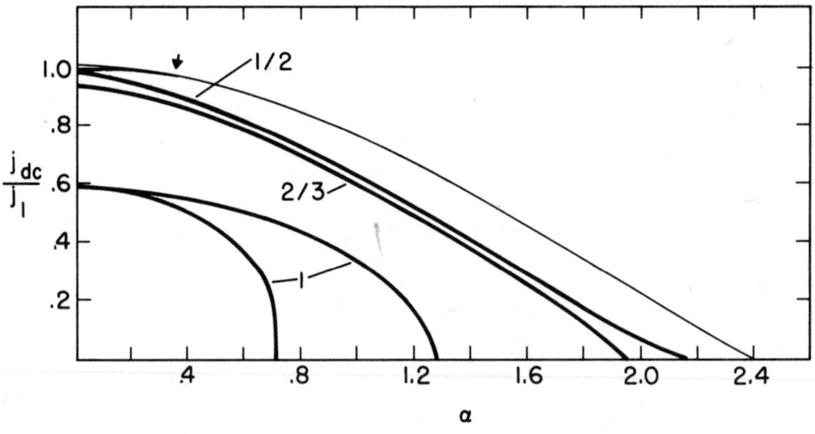

Figure 1. Normalized dc current versus normalized rf voltage α for some of the resonances for $\omega_p/\omega_J = 0.9$ as predicted by the pseudo-voltage-bias model.[17] α is defined as $2\gamma H_{rf}/\omega$, where ω is the angular frequency. Two solutions for the fundamental (1) and the 1/2 fractional harmonic and one solution for the 2/3 fractional harmonic are shown in addition to the maximum current (fine line). An arrow indicates the point at which the upper (1/2) branch terminates on the maximum current curve.

A study of some of these analog effects in superfluid ^3He will, it is hoped, lead to a deeper understanding of the longitudinal nuclear magnetic resonance and to the superfluid state in general.

References

(1) D. D. Osheroff and W. F. Brinkman, Phys. Rev. Lett. 32, 584 (1974).

(2) H. M. Bozler, M. E. R. Bernier, W. J. Gully, R. C. Richardson, and D. M. Lee, Phys. Rev. Lett. 32, 875 (1974).

(3) R. A. Webb, R. L. Kleinberg, and J. C. Wheatley, Phys. Rev. Lett. 33, 145 (1974); Phys. Lett. 48A, 421 (1974).

(4) A. J. Leggett, Phys. Rev. Lett. 29, 1227 (1972); 31, 352 (1973); J. Phys. C: Proc. Phys. Soc. (London) 6, 3187 (1973).

(5) A. J. Leggett, Ann. Phys. (N.Y.) 85, 11 (1974).

(6) K. Maki and T. Tsuneto, Prog. Theor. Phys. 52, 773 (1974).

(7) B. D. Josephson, Phys. Lett. 1, 251 (1962).

(8) N. F. Pedersen, T. F. Finnegan, and D. N. Langenberg, Phys. Rev. B6, 4151 (1972).

(9) D. N. Langenberg, Rev. de Phys. Appl. (Suppl. J. de Phys.) 9, 35 (1974).

(10) R. E. Harris, Phys. Rev. B10, 84 (1974), and references therein.

(11) P. W. Anderson, in Prog. Low Temp. Phys., C. J. Gorter, ed. (North-Holland, Amsterdam, 1967), Vol. V, p. 1.

(12) D. B. Sullivan and J. E. Zimmerman, Am. J. Phys. 39, 1504 (1971).

(13) G. I. Rochlin and P. K. Hansma, Am. J. Phys. 41, 878 (1973).

(14) A. J. Dahm, A. Denenstein, T. F. Finnegan, D. N. Langenberg, and D. J. Scalapino, Phys. Rev. Lett. 20, 859 (1968); 20, 1020E (1968).

(15) N. R. Werthamer, Phys. Rev. 147, 255 (1966).

(16) Maki has calculated this term to be very small if it exists (private communication).

(17) A. J. Dahm and D. N. Langenberg, J. Low Temp. Phys. 19, 145 (1975).

(18) O. P. Balkashin, L. I. Ostrovsky, and I. K. Yanson, Phys. of the Condensed State, No. 28, 3 Khar'kov (1973).

(19) C. K. Bak, B. Kofoed, N. F. Pedersen, and K. Saermark (to be published in Appl. Phys. Lett.).

FREE ENERGY FUNCTIONALS FOR SUPERFLUID ^{3}He

J. W. Serene

Institute of Theoretical Physics, Department of Physics
Stanford University, Stanford, California 94305

and

D. Rainer

Laboratory of Atomic and Solid State Physics
Cornell University, Ithaca, New York 14850

From the first suggestions in the early 1960's of the possibility of pairing in ^{3}He, until well after the discovery of the A and B phases in 1972, theoretical investigations of the properties of ^{3}He with pairing took as a working assumption that this system would be well described by the simplest possible combination of BCS pairing theory and Landau's Fermi liquid theory. The Fermi liquid theory provides an exact account of normal ^{3}He just above T_c in terms of interacting quasi-particles, and corrections to BCS pairing theory for these quasi-particles were expected to be of order $(k_F\xi_0)^{-1} \sim (T_c/T_F) \sim 3 \times 10^{-3}$. However, experiments on the A and B phases soon showed this assumption to be incorrect.[1] In particular, the experiments imply that both new phases have odd-ℓ, spin-triplet pairing; that ℓ is the same in both phases; and that ^{3}He-A is an equal-spin pairing state, while ^{3}He-B is not. BCS pairing theory, on the other hand, predicts that the equilibrium state for triplet pairing is never an equal-spin pairing state.[2] Furthermore, in simple BCS theories $\Delta C/C_N$, the ratio of the specific heat discontinuity at T_c to the specific heat in the normal phase just above T_c, is always ≤ 1.43, while experiments at melting pressure give $\Delta C/C_N \cong 2.0$.

One possible response to the failures of simple BCS pairing theory is to assume that the "strong-coupling" corrections to the theory act only to stabilize the correct phases and to change the overall amplitude and temperature dependence of the order parameter. With this scheme one calculates properties of interest exactly as in BCS theory, but at the end of the calculation treats the order parameter as an unknown to be determined from experiment. While this scheme may be adequate for some properties, such as the superfluid density or zero sound attenuation, it surely fails whenever the relative energies of different states are relevant. Important problems which fall in the latter category include the equilibrium states in confined geometries and the structure of vortex cores. For example, we show in Fig. 1 a conjectured phase diagram for ^{3}He below the polycritical point in a long cylinder of diameter d. The details of this diagram depend sensitively on the strong-coupling corrections to the energies of the various possible states.

A second possibility is to construct a phenomenological theory based on symmetries alone, using the BCS theory only as a guide to the order of magnitude of the open parameters. Two outstanding examples of this approach are Leggett's theory of NMR[3] and Mermin and Stare's analysis of the $\ell = 1$ models for the A and B phases.[4] Unfortunately this approach becomes unwieldy if the number of relevant open parameters is too large. For example, the general fourth order $\ell = 1$ free energy functional for the spatially non-uniform case is:

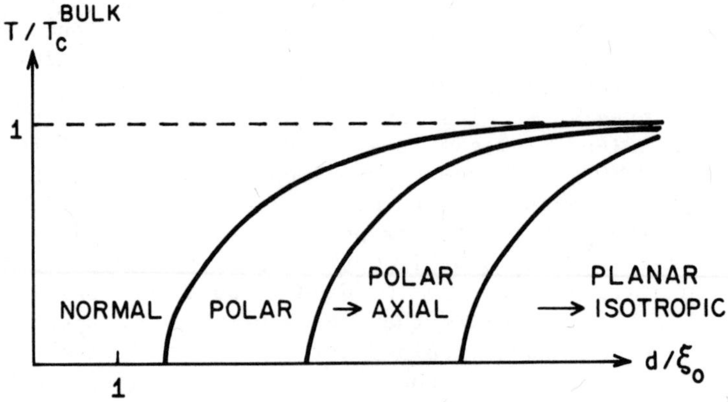

Figure 1. Conjectured Phase diagram in long circular cylinder, diameter = d.

$$F_S - F_N = \int d^3r\{\frac{T-T_c}{T_c} a\, Tr(A(\vec{r})A^+(\vec{r})) + \beta_1|Tr(A(\vec{r})\tilde{A}(\vec{r}))|^2$$

$$+ \beta_2[Tr(A(\vec{r})A^+(\vec{r}))]^2 + \beta_3 Tr(A(\vec{r})\tilde{A}(\vec{r}))(A(\vec{r})\tilde{A}(\vec{r}))^*$$

$$+ \beta_4 Tr(A(\vec{r})A^+(\vec{r}))^2 + \beta_5 Tr(A(\vec{r})A^+(\vec{r}))(A(\vec{r})A^+(\vec{r}))^*$$

$$+ \frac{1}{3} N(0)\xi_T{}^2(\lambda|\vec{\nabla}\cdot\vec{A}_\alpha(\vec{r})|^2 + |\vec{\nabla}\times\vec{A}_\alpha(\vec{r})|^2)\} \tag{1}$$

which contains the eight parameters a, $\beta_1 - \beta_5$, ξ_T^2, and λ. Specific heat discontinuity measurements in magnetic fields can determine at most three (only two above the PCP) independent linear combinations of $\beta_1/a,\ldots,\beta_5/a$, while Mermin and Stare's analysis of the functional without gradient terms only restricts the β_i to certain regions in the 5-dimensional β-space. Near T_c the BCS results $\xi_T^2 = (7/20)\zeta(3)\xi_0^2$, $\lambda = 3$ are exact, but at lower temperatures strong-coupling corrections to these coefficients may also be important. Thus the empirically accessible information is not sufficient to determine the results of calculations on size effects and related problems. Because these problems must be attacked numerically, uncertainty about the coefficients represents a particularly serious handicap.

The best way out of this difficulty would be to calculate the coefficients in the phenomenological free energy functional directly from a microscopic theory. The first microscopic calculations of strong-coupling corrections to the β_i were done in spin-fluctuation theory.[6] These calculations provided both an understanding of the origin of the corrections to weak-coupling theory, and a qualitative explanation of the shape of the equilibrium phase diagram. However, one cannot trust a one parameter model such as spin-fluctuation theory to give quantitatively accurate results for ^{3}He.

The BCS weak-coupling free energy functional is the leading term, of order $(T_c/T_F)^2$, in an expansion of the free energy difference $\Delta F = F_S - F_N$ in powers of T_c/T_F. In the BCS free energy one needs to know two parameters: the transition temperature T_c and the normal quasi-particle density of states N(0). We have shown that the leading strong-coupling corrections to the BCS free energy are of order $(T_c/T_F)^3$ and can be calculated from the 2-particle scattering amplitude for quasi-particles on the Fermi surface in the normal state. The same scattering amplitude determines the normal state transport coefficients, so in this sense the strong-coupling free energy can be calculated from Fermi liquid theory.

To obtain this result we begin from a stationary free energy functional of the type first introduced by Luttinger and Ward.[7,8] We use this formalism both because the stationarity property simplifies our calculations, and also because the formalism can be extended in a straightforward way to more complicated problems. In a 4 × 4 matrix notation the free energy as a functional of the self-energy $\hat{\Sigma}(\vec{k},\omega_n)$ is

$$F\{\hat{\Sigma}\} = -\frac{1}{2} k_B T \sum_{\omega_n} \int \frac{d^3k}{(2\pi)^3} \mathrm{Tr}\{\hat{\Sigma}(\vec{k},\omega_n)\hat{G}(\vec{k},\omega_n) + \log(-\hat{G}(\vec{k},\omega_n)^{-1})\} + \phi\{\hat{G}\} \quad , \tag{2}$$

where $\hat{G}$ is related to $\hat{\Sigma}$ by the Dyson equation

$$\hat{G}^{-1} = \hat{G}_0^{-1} - \hat{\Sigma} \quad , \tag{3}$$

and $\phi\{\hat{G}\}$ is a functional represented by a set of diagrams without external lines. $\hat{\Sigma}$ and $\hat{G}$ are defined in terms of the more familiar 2 × 2 self-energies and Green's functions by

$$\hat{\Sigma}(\vec{k},\omega_n) = \begin{bmatrix} \Sigma(\vec{k},\omega_n) & \Delta(\vec{k},\omega_n) \\ \Delta^+(\vec{k},-\omega_n) & -\Sigma^T(-\vec{k},-\omega_n) \end{bmatrix}$$

$$\hat{G}(\vec{k},\omega_n) = \begin{bmatrix} G(\vec{k},\omega_n) & F(\vec{k},\omega_n) \\ F^+(\vec{k},-\omega_n) & -G^T(-\vec{k},-\omega_n) \end{bmatrix} \quad , \tag{4}$$

and the stationarity condition is that

$$\frac{\delta F\{\hat{\Sigma}\}}{\delta\hat{\Sigma}(\vec{k},\omega_n)} = 0 \quad \text{if} \quad \frac{\delta\phi\{\hat{G}\}}{\delta\hat{G}(\vec{k},\omega_n)} = \frac{1}{2}\hat{\Sigma}^T(\vec{k},\omega_n) \quad . \tag{5}$$

The physical self-energy is always a stationary point at which $F\{\hat{\Sigma}\}$ is equal to the actual thermodynamic free energy, so we expand $F\{\hat{\Sigma}\}$ around the normal self-energy $\hat{\Sigma}_N$ and Green's function $\hat{G}_N$, and identify the coefficients of the terms of fourth order in Δ with the β_i. Because of the self-consistency condition, equation (5), we only need to expand $\Delta\phi = \phi\{\hat{G}\} - \phi\,\hat{G}_N$.

Our expansion of $\Delta\phi$ is based on three assumptions:

1) T_c is small compared to all relevant characteristic normal

state energies, and the smallest of these energies is of order T_F; the strong-coupling corrections to the free energy can be calculated adequately to leading order in T_c/T_F.

2) For $|\omega_n| \lesssim k_B T_F$ and $|k-k_F| \lesssim k_F$, the normal state Green's function consists of a quasi-particle part and an incoherent part,

$$G_N(\vec{k},\omega_n) = G_N(\vec{k},\omega_n)_{qP} + G_N(\vec{k},\omega_n)_{inc} , \qquad (6)$$

where $G_N(\vec{k},\omega_n)_{inc} \sim (k_B T_F)^{-1}$.

3) The scale of the normal state vertex functions is set by k_F and $k_B T_F$; in particular these functions do not introduce factors of (T_F/T_c) in the expansion of $\Delta\phi$. The 4-point normal vertex function varies with frequency and energy on the scale of $k_B T_F$. We emphasize that these assumptions, which have been made at least implicitly in all spin-fluctuation model free energy calculations, are actually sufficient to derive a much more general result for the free energy functional.

Using assumptions (1)-(3) we can classify the diagrams for $\Delta\phi$ by their order in T_c/T_F. Fig. 2 shows the weak-coupling diagram, of order $(T_c/T_F)^2$ and all the corrections of order $(T_c/T_F)^3$. The weak-coupling diagram, Fig. 2a, always leads to β's with the same ratios as found from BCS theory, but the overall magnitude of the β's and the coefficient a of the Δ^2 term can differ from BCS. We do not know how to calculate these corrections accurately, but have good reasons to believe that at least the corrections to the β's are small. The remaining diagrams in Fig. 2 are of order $(T_c/T_F)^3$ and represent strong-coupling corrections which change the ratios of the β_i. To order $(T_c/T_F)^3$, only the quasi-particle Green's functions contribute to these diagrams, and the normal state vertex functions can be evaluated at zero frequency with all four momenta on the Fermi surface. All remaining quasi-particle renormalization factors can then be absorbed by replacing these vertex functions with the dimensionless renormalized quasi-particle scattering amplitude $T(\theta\ \phi)$. The frequency sums and integrals over the energies in the quasi-particle Green's functions are straightforward, and the strong-coupling contributions of order $(T_c/T_F)^3$ can finally be expressed in terms of angular averages of the form

$$\langle w(\theta,\phi)T^{(\alpha)}(\theta,\phi)T^{(\beta)}(\theta,\phi)\rangle \qquad (7a)$$

or

$$\langle w(\theta,\phi)T^{(\alpha)}(\theta,\phi)T^{(\beta)}(\theta',\phi')\rangle \qquad (7b)$$

where α and β label the spin symmetric or spin antisymmetric scattering amplitudes. Integrals of the form (7b) in which the two

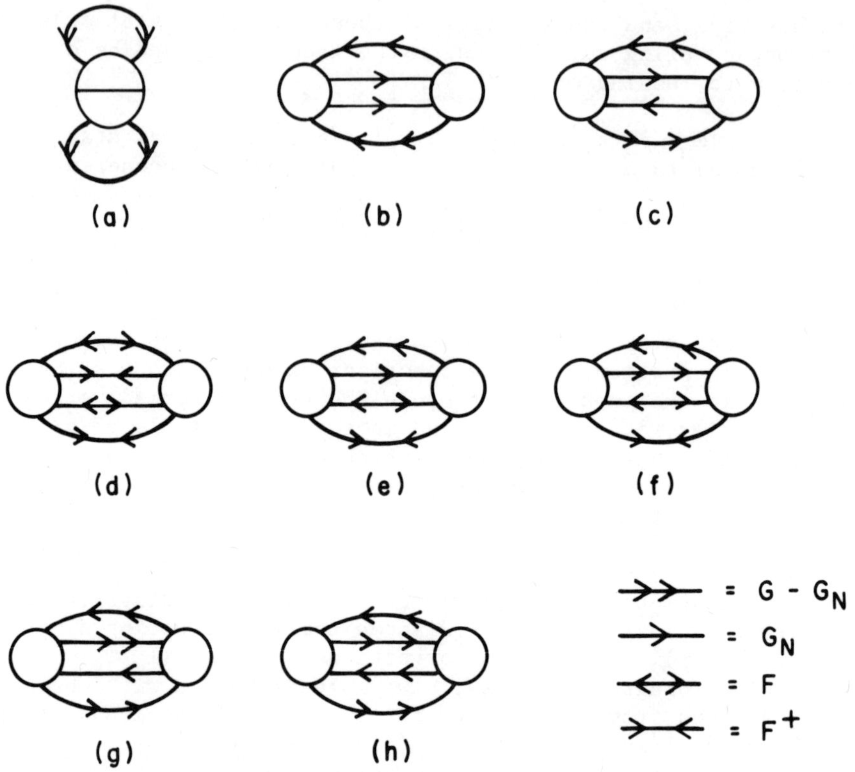

Figure 2. (a) The weak coupling diagram for $\Delta\phi$. (b) - (h) All corrections to $\Delta\phi$ of order (T_c/T_F) .

scattering amplitudes are evaluated at different points come from Figures 2(e) and 2(f); all the other diagrams give integrals of the type (7a). These results hold for all $T \leq T_c$ and for any ℓ. The free energy depends on ℓ through the weighting factors $w(\theta,\phi)$ which come from the angular dependence of the order parameter.

Thus, to calculate the strong-coupling free energy one must approximate the quasi-particle scattering amplitude $T(\theta,\phi)$. The spin-fluctuation model free energy results from a simple one parameter approximation for T. If we used the spin-fluctuation theory scattering amplitude, our calculation would reduce to that of Brinkman, Serene, and Anderson[6] (BSA), with the following modifications. First, BSA evaluated the susceptibilities in the small q limit, and inserted a large q cutoff in their q-integrals; we make no such approximations (at the price of more complicated angular integrals). Second, our calculation includes spin-fluctuation theory diagrams of the type shown in Fig. 3(a), which BSA omitted; these diagrams are negligible only in the very strong enhancement limit.

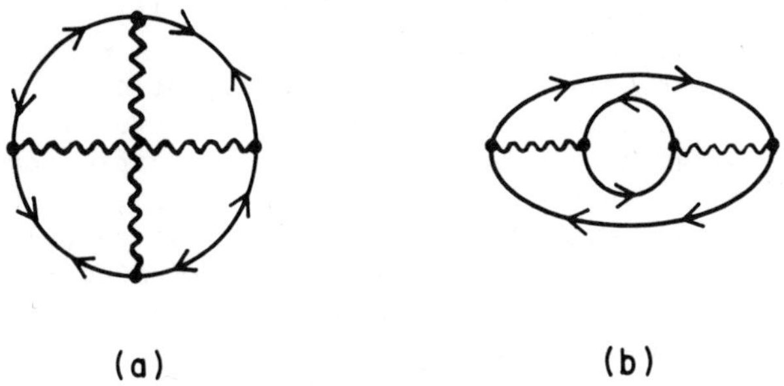

Figure 3. Diagrams omitted in the work of Ref. 6.

Finally, BSA dropped the diagram shown in Fig. 3(b), which is included correctly in our calculation.

We have calculated the strong-coupling free energy functional for $\ell = 1$ using for the scattering amplitude the s-p approximation introduced by Dy and Pethick.[9] At low pressures this approximation gives excellent agreement with the measured transport coefficients. We use the Landau parameters F_0^s, F_0^a, and F_1^s given by Wheatley,[1] put $F_\ell = 0$ for $\ell \geq 2$, and fix F_1^a, by the forward scattering sum rule (our results are insensitive to F_1^a). If we take the BCS values for the weak-coupling β_i, then we find, in agreement with experiment, that the isotropic state is stable at low pressures, and the axial state is stable at high pressures. We calculate the crossover between the isotropic and axial states to be at approximately 9 bar, compared to the experimental polycritical pressure of approximately 21.5 bar. These results are encouraging, since they require calculating a small free energy difference. The results for the specific heat discontinuity are less satisfactory. At zero pressure we find $\Delta C/C_N = 1.53$, which is probably only slightly too large. However, at melting pressure our calculation gives $\Delta C/C_N = 6.62$, while experimentally $\Delta C/C_N \cong 2.0$. The disagreement is slightly less glaring when expressed in terms of the β's: the experiments imply $\Delta\beta_2 + \Delta\beta_4 + \Delta\beta_5 = -0.81|\beta_1^{BCS}|$ $(\Delta\beta_i = \beta_i - \beta_i^{BCS})$, and theory gives $\Delta\beta_2 + \Delta\beta_4 + \Delta\beta_5 = -1.64|\beta_1^{BCS}|$.

We have also calculated the $\ell = 3$ free energy in the same approximation. The results do not stabilize any of the $\ell = 3$ states hoped to be consistent with the experimental properties of superfluid ^{3}He.

To conclude, we will discuss briefly three possible sources of the disagreement between theory and experiment at high pressures. First, the s-p approximation may well be inadequate at high pressures. The viscosity at melting pressure calculated in the s-p approximation is smaller than the measured viscosity by approximately a factor of 2, which implies that the appropriate angular average of T^2 is too large by a factor of 2. It is at least suggestive that the calculated corrections to $\beta_2 + \beta_4 + \beta_5$ at melting pressure are also too large by a factor of 2.[10] A second possibility is that corrections to BCS theory are important in the weak-coupling diagram. In particular, the specific heat discontinuity is proportional to the coefficient of Δ^2, which comes solely from the weak-coupling diagram. We have calculated this coefficient in spin-fluctuation theory, and find corrections to BCS of only about 5 percent. Finally, it may be necessary to include in the free energy terms of order (T_c/T_F).[4] In this case there is no hope for a quantitatively accurate microscopic calculation, because to this order one needs the incoherent parts of the Green's functions and also the normal state four, six, and eight-point vertex functions at large frequencies and far from the Fermi surface. Our hope is that the first explanation is correct. Since the s-p approximation works well for transport coefficients at low pressures, careful measurements of the B, A_1, and A_2 specific heat discontinuities at low pressures would test this possibility.

References

(1) Further details and original references on the experimental picture of superfluid ^{3}He can be found in J. C. Wheatley, Rev. Mod. Phys. 47, 415 (1975).

(2) R. Balian and N. R. Werthamer, Phys. Rev. 131, 1553 (1963).

(3) A. J. Leggett, Ann. Phys. (N.Y.) 85, 11 (1974).

(4) N. D. Mermin and G. Stare, Phys. Rev. Lett. 30, 1135 (1973); G. Stare, 1974, Thesis, Cornell University, unpublished.

(5) V. Ambegaokar, P. G. de Gennes, and D. Rainer, Phys. Rev. A 9, 1676 (1974).

(6) W. F. Brinkman, J. W. Serene, and P. W. Anderson, Phys. Rev. A 10, 1386 (1974).

(7) J. M. Luttinger and J. C. Ward, Phys. Rev. 118, 1417 (1960).

(8) C. DeDominicis and P. C. Martin, J. Math. Phys. 5, 31 (1964).

(9) K. S. Dy and C. J. Pethick, Phys. Rev. 185, 373 (1969).

(10) C. J. Pethick, H. Smith, and P. Bhattacharyya, Phys. Rev. Lett. 34, 643 (1975).

FLUCTUATIONS ABOVE THE SUPERFLUID TRANSITION IN LIQUID ^{3}He*

V. J. Emery

Brookhaven National Laboratory

Upton, New York 11973

Abstract: It is shown that fluctuations above the superfluid transition in liquid ^{3}He depend strongly upon the relative angular momentum ℓ of a Cooper pair but are insensitive to the fourth order term in the Ginsburg-Landau free energy. The effects are shown to be observable in the static magnetization, viscosity and spin diffusion and give a means of determining the value of ℓ.

I. INTRODUCTION

There is a considerable amount of evidence to support the idea that superfluidity in liquid ^{3}He is produced by Cooper pairing of quasi-particles in a state of relative angular momentum ℓ equal to one. Nevertheless, at present, there are several discrepancies between theory and experiment and, whenever they arise, it is hard to know if they should be attributed to the experiments, the background theory or the assumed value of ℓ. There is no doubt that ℓ is odd, the question is whether or not it might turn out to be 3 or 5 etc., rather than 1.

The difficulty in establishing this with certainty may be seen by considering the situation just below the normal-superfluid transition temperature T_c where the difference F between the superfluid and normal free energies per unit volume may be expanded in powers of the order parameters Δ_α:

---

*Work supported by Energy Research and Development Administration.

$$F = -\frac{N_0}{2}\theta\sum_\alpha |\Delta_\alpha|^2 + F_4(\Delta_\alpha)\ldots \tag{1}$$

where N_0 is the density of states of one spin at the fermi surface, $\theta = (T-T_c)/T_c$ and F_4 is of fourth order in the Δ_α. The equilibrium values of Δ_α are found by minimizing F. For any ℓ there are $3(2\ell+1)$ complex order parameters (one for each of the $(2\ell+1)$ projections of angular momentum for each component of the spin triplet) and the complexity of minimizing F increases rapidly with ℓ. The form of F_4 may be simplified by imposing the general conditions of gauge invariance and invariance under separate rotations in spin-space and real space,[1] but it always involves several parameters which are not well-known because weak-coupling BCS theory is not applicable.[1] Thus the problem involves a classification of minima in the various regions of a many-parameter space and this has not been carried out with any semblance of completeness except for $\ell = 1$. The pragmatic approach therefore has been to try to fit experiments with $\ell = 1$ until a discrepancy arises, and there have been very few calculations for higher ℓ. The purpose of this note is to point out that many of these problems may be avoided by studying the effects of fluctuations of the order parameter just above T_c, since they are very sensitive to the value of ℓ yet do not depend strongly upon F_4. It will be shown that measurement of the static magnetization, viscosity or spin diffusion close to T_c can give direct information about ℓ.

Fluctuations in the equilibrium properties of clean superconductors were first studied by Thouless[2] and transport coefficients were discussed by Emery.[3] It was found that the long coherence length gives rise to a very narrow fluctuation region for the specific heat but the effects are observable in transport coefficients. Subsequently, it was pointed out that fluctuations are considerably enhanced in dirty superconducting films and this led to a great deal of both theoretical and experimental activity verifying the observation of their effects.[4] This way of enhancing fluctuations is not open to us at present in superfluid ^{3}He, but it will be shown that they should, nevertheless be observable. At first sight it seems more profitable to study transport properties but a measurement of the static magnetization of liquid ^{3}He by Paulson, Kojima and Wheatley[5] is capable of such accuracy that it may well have seen fluctuation effects already and accordingly this possibility will be discussed first.

II. EQUILIBRIUM PROPERTIES

To calculate equilibrium properties[6] above T_c, F in Eq. (1) is regarded as a function of position $\underset{\sim}{R}$ and a term $\frac{N_0}{2}\sum_\alpha \xi_\alpha^{\,2}\left|\frac{\partial\Delta_\alpha}{\partial \underset{\sim}{R}}\right|^2$ is added to take account of the increase in energy for a varying order parameter. If the new free energy density is $F'(\tilde{R})$, the partition function is obtained by integrating $\exp(-\beta\int d\tilde{R}\, F'(\tilde{R}))$ over all functions $\Delta_\alpha(\tilde{R})$ The ξ_α are coherence lengths for the various Δ_α. The importance of F_4 may be estimated from mean field theory. Apart from a shift in T_c which is unimportant because T_c is a parameter of the theory, the correction is of order $\sum_\alpha \xi_\alpha^{-3}\,\theta^{-1/2}$. Since the ξ_α are quite long in a fermion superfluid transition, F_4 is unimportant until θ is very small and in fact, for superfluid ^{3}He, it does not affect critical exponents until $\theta \simeq 10^{-6}$. For practical purposes, then, it is sufficient to ignore F_4 and hence to consider only Gaussian fluctuations.

In order to carry out explicit calculations, it is necessary to know the ξ_α, which are assumed to be given in the Ginsburg-Landau theory considered here, but may be found from a microscopic calculation. It may be shown[7] that the ξ_α depend only upon ℓ and m and that they may be written

$$(\xi_{\ell m}/\xi_{00})^2 = \frac{\int d\hat{k}\,|Y_\ell^{\,m}(\hat{k})|^2(\hat{k}\cdot\hat{K})^2}{\int d\hat{k}\,|Y_0^{\,0}(\hat{k})|^2(\hat{k}\cdot\hat{K})^2} \tag{2}$$

where ξ_{00} is the usual s-state coherence length given by

$$(k_F\xi_{00})^2 = \frac{7\zeta(3)}{12\pi^2}\left(\frac{T_F}{T_c}\right)^2 \tag{3}$$

with $\zeta(3)$ Riemann's zeta function of argument 3. Equation (2) comes about because $\tilde{R}$ in the free energy density is the center of mass of a Cooper pair and for total momentum $\tilde{K}$, the most important contribution to the square gradient term is of the form $(\tilde{K}\cdot\hat{k})^2$ averaged over the wave function of a pair which has components proportional to $Y_\ell^{\,m}(\tilde{k})$. ($\tilde{k}$ is the relative momentum of a pair.)

The integral in Eq. (2) may be evaluated to give

$$\left(\frac{\xi_{\ell m}}{\xi_{00}}\right)^2 = \frac{3}{2}\left[1 + \frac{1-4m^2}{(2\ell+3)(2\ell-1)}\right] \tag{4}$$

For example, for $\ell = 1$, $(\xi_{10}/\xi_{00})^2 = 9/5$ and $(\xi_{11}/\xi_{00})^2 = 3/5$ in agreement with the result of Ambegaokar, de Gennes and Rainer.[8]

Consider now the static magnetization, which is a constant in the normal state and the A phase, with a very slight change in value at T_c. Paulson, Kojima and Wheatley[5] found that the change from the normal value to the A-phase value had an unexpected rounding. This is partly a consequence of thermal inhomogeneity but it does not seem possible to explain the whole effect in this way, and it was pointed out by Patton[10] that fluctuations may be responsible. He assumed $\ell = 1$ and could only get a large enough effect if he reduced the value of $\xi_{00}{}^3$ by a factor between 5 and 15. However, it seems that ξ_{00} is not an adjustable parameter and that the freedom to fit the experiments is gained by allowing $\ell \neq 1$. It is straightforward to extend Patton's calculation to arbitrary ℓ and find that the fluctuation contribution to the magnetization satisfies

$$\delta M \propto \xi_{00}{}^{-3} \sum_{m=-\ell}^{\ell} (\xi_{00}/\xi_{\ell m})^3 \tag{5}$$

From Eq. (4) it may be seen that the coefficient of $\xi_{00}{}^{-3}$ in Eq. (5) is 4.7 for $\ell = 1$, 14.1 for $\ell = 3$ and 26.4 for $\ell = 5$. First of all, this shows the sensitivity of δM to the value of ℓ and secondly that δM can be increased by a factor of 5 or more over the $\ell = 1$ value by having $\ell \geq 5$. However, it should be realized that the experiments were not designed to measure the fluctuation effect and that the result is uncertain because of a difficulty in determining the baseline with sufficient accuracy.[11] Accordingly the conclusion which should be drawn from this is not that $\ell \geq 5$ but rather that the experiment can potentially determine ℓ (or at least bound it) and is worth repeating with this aim in view.

III. TRANSPORT PROPERTIES

To evaluate transport coefficients, it is simplest to use the Boltzmann equation for which the solution is well-developed for fermi liquids.[12,13] The method used in Ref. 3 needs some modification since it made use of Abrikosov and Khalatnikov's solution[14] of the Boltzmann equation which assumes that the total momentum $\underset{\sim}{K}$ of a pair of quasi-particles satisfies $K \gg k_F T/T_F$, whereas, for a discussion of fluctuation effects, the opposite limit is relevant. Also it is not possible to assume that the scattering particles are on the fermi surface. Details of the solution of the Boltzmann equation are given in a separate publication[7] and only a summary of the physical effects and main conclusions will be given here.

At T_c, two quasi-particles on the fermi surface with total momentum zero have a divergent scattering amplitude and above T_c, the impending transition manifests itself through a near divergence. This has no effect on the thermal conductivity since particles of total momentum zero do not contribute to the heat current. For the viscosity η and the spin-diffusion coefficient D, the scattering rates are weighted averages of the square modulus of the scattering amplitude A which is given by

$$A = \frac{4\pi}{N_0} \sum_{m=-\ell}^{\ell} \frac{Y_\ell^m(\hat{\underset{\sim}{k}})Y_\ell^{-m}(\hat{\underset{\sim}{k}})}{\theta + K^2\xi_{\ell m}^2 + i\pi\beta\varepsilon/4} \tag{6}$$

where N_0 is the density of states at the fermi surface, ε is the incident energy and $\hat{\underset{\sim}{K}}$ is the pole of the polar coordinates of $\hat{\underset{\sim}{k}}$. As before, $\underset{\sim}{K}$ and $\underset{\sim}{k}$ are respectively the total and relative momentum of a pair of quasi-particles. The scattering rate has to be integrated over $\underset{\sim}{K}$ and ε and it is therefore finite at $\theta = 0$. The results of the calculation of the viscosity will be given in more detail since the fluctuation effect is larger than for the spin diffusion coefficient and there are experiments already available. For $\ell = 1$, the change in the mean free time is given by

$$\delta\tau \simeq \frac{2.43T_F}{(k_F\xi_{00})^3}(3\tau_A + \tau_{QP})^2(\lambda - \theta^{1/2}\tan^{-1}\frac{\lambda}{\theta^{1/2}}) \tag{7}$$

Here we have used $\hbar = 1$, τ_A is the viscous lifetime calculated by Abrikosov and Khalatnikov[14] and τ_{QP} is the lifetime of a quasi-particle on the fermi surface. The integral over $\underset{\sim}{k}$ was cut off at $\lambda\xi_{\ell\ell}^{-1}$ and the numerical factor comes from a rather complicated function of the $\xi_{\ell m}$. The result for $\ell = 3$ is obtained by multiplying by 2.75. Equation (7) was obtained by using an approximate solution of the Boltzmann equation,[13] assuming $\delta\tau$ small. In the same approximation, the fermi liquid mean free time is[13]

$$\tau_0 = \frac{\pi^2}{12}\tau_{QP} + \frac{3}{4}(\tau_A - \tau_{QP}) \tag{8}$$

The value of $\delta\tau$ is uncertain because the experimental value of τ_0 does not determine both τ_A and τ_{QP} and because the cutoff λ has to be estimated. The uncertainty is limited by the inequality $0 \leqq \tau_{QP}/\tau_A \leqq 3$. The value of λ is determined by factors which arise in the solution of the Boltzmann equation and it is about 0.5. Notice that the coefficient of $\theta^{1/2}$ does not depend upon λ as $\theta \to 0$. Near to the melting curve, $k_F T_F \tau_0/\hbar = 1.3 \times 10^4$ and $(k_F\xi_{00}) = 105$ so Eq. (7) becomes

$$\frac{\delta\tau}{\tau_0} \simeq \gamma(1 - 2\theta^{1/2}\tan^{-1}\frac{1}{2\theta^{1/2}}) \tag{9}$$

where the inequality for τ_{QP}/τ_A gives $0.2 \leqq \gamma \leqq 0.48$.

Alvesalo *et al.*[15] have measured the viscosity along the melting curve and there are no signs of fluctuation effects in their data down to 2.63 mK. Using $T_c = 2.57$ mK, the smallest value of θ is

0.023 and, according to Eq. (9) $\delta\tau/\tau_0 \simeq 0.6\gamma$ so that the overall effect is bigger than about 10%. Given the uncertainty in λ and the fact that the experiment was not designed to look for fluctuations, it is perhaps not unreasonable that they were not seen if $\ell = 1$, and again it is suggested that the experiment should be repeated with this aim in view. On the other hand, it is difficult to understand how the effect could have been missed if $\ell = 3$ since it is 2.75 times as large as for $\ell = 1$. This may perhaps be taken as some evidence in favor of $\ell = 1$ although the observation of a decrease in τ would clearly make this much more definite.

References

(1) N. D. Mermin and C. Stare, Phys. Rev. Lett. 30, 1135 (1973).

(2) D. J. Thouless, Ann. Phys. (N.Y.) 10, 533 (1960).

(3) V. J. Emergy, Ann. Phys. (N.Y.) 28, 1 (1964).

(4) For a recent review and list of references see M. Tinkham, Low Temperature Physics LT13, eds. K. D. Timmerhaus, W.-J. O'Sullivan and E. F. Hammel (Plenum Press, New York, 1974).

(5) D. N. Paulson, H. Kojima and J. C. Wheatley, Phys. Lett. 47A, 457 (1974).

(6) T. M. Rice, Phys. Rev. 140, A1889 (1965).

(7) V. J. Emery, to be published.

(8) V. Ambegaokar, P. G. de Gennes and D. Rainer, Phys. Rev. A9, 2676 (1974).

(9) S. Takagi, Prog. Theor. Phys. 51, 1998 (1974).

(10) B. R. Patton, Phys. Lett. 47A, 459 (1974).

(11) J. C. Wheatley, private communication.

(12) G. A. Brooker and J. Sykes, Phys. Rev. Lett. 21, 279 (1968); H. H. Jensen, H. Smith and J. W. Wilkins, Phys. Lett. 27A, 532 (1968).

(13) V. J. Emery and D. Cheng, Phys. Rev. Lett. 21, 533 (1968); V. J. Emery, Phys. Rev. 175, 251 (1968).

(14) A. A. Abrikosov and I. M. Khalatnikov, Repts. Progr. Phys. 22, 329 (1959).

(15) T. A. Alvesalo, H. K. Collan, M. T. Loponen, O. V. Lounasmaa and M. C. Veuro, J. Low Temp. Phys. 19, 1 (1975).

EXCITATIONS IN DILUTE ^3He-^4He MIXTURES

Laurence Mittag*

Center for Theoretical Studies, University of Miami
Coral Gables, Florida 33124

and

Michael J. Stephen†

Physics Department, Rutgers University
New Brunswick, New Jersey 08903

Abstract: The energy spectrum of elementary excitations in dilute He^3-He^4 mixtures is investigated using a Feynman type of wave function. It is found that the He^3 energy spectrum lies below the phonon-roton energy spectrum and exhibits a minimum near the roton minimum. Reinterpretation of their experimental results by Sobolev and Esel'son confirms this description of the He^3 spectrum. Some consequences of this spectrum are discussed.

The two-roton Raman spectra of superfluid He^3-He^4 mixtures have been measured by Surko and Slusher[1] at a temperature of 1.3K for He^3 molar concentrations less than 31% and by Woerner, Rockwell and Greytak[2] at a temperature of 0.60K for concentrations less than 11%. The roton energy Δ_4 determined by these experiments is essentially independent of the He^3 concentration X, at least for X less than 31%. (The linewidth increases with increasing concentration.) It is also possible to obtain the roton energy in mixtures by fitting the formula $\rho_n = n_3 m_3^* + \rho_{nr} + \rho_{np}$ to experimental normal fluid density data[3], where n_3 is the He^3 number density, m_3^*

*Supported in part by the National Institute of Health.

†Supported in part by the National Science Foundation.

is the He^3 effective mass and the last two terms are the roton and phonon contributions. The roton energy obtained in this way decreases markedly with concentration, e.g., $\Delta_4 \simeq 5.1K$ at $X = .30$, in striking contrast to the Raman-scattering results. L. P. Pitaevsky[4] has suggested that this dilemma could be resolved if the energy spectrum of the He^3 excitations in dilute mixtures exhibited a minimum similar to the roton minimum in pure He^4.

There is a simple theoretical argument which indicates that Pitaevsky is probably correct: Let us assume that He^3 and He^4 have the same mass and potential energy, i.e., $V_{34} = V_{44}$. If we now construct a wave function for a localized excitation (roton) in the mixture in the manner of Feynman[5] and Cohen[6] and let one atom in the excitation be He^3 then the condition that the wave function be totally symmetric in all its arguments can be relaxed. This results in a larger class of wave functions with which to minimize the energy. Presumably this will lead to a lower energy. At low wave numbers, the He^3 excitation energy will have the form $\hbar^2k^2/2m_3^*$ and will lie below the phonon branch. At larger wave numbers, the argument above indicates that the energy of He^3 excitations will probably lie below the rotons and exhibit a minimum also. After all, the roton minimum reflects the first maximum in the structure factor for $k \simeq (2\pi)/r_0$ where r_0 is the interparticle separation. Since the He^3 molar volume exceeds that of He^4 by about 30%, r_{34} will exceed $r_{44} = r_0$ by about 10% and k_3 the wave number at the minimum will be reduced accordingly. The argument above is not applicable to other impurities, e.g., electrons or protons in He^4 because of the different masses and potentials.

Suppose a He^3 atom is dissolved in superfluid Helium. The ground-state wave function is $\Phi(\bar{R},\bar{r}_1,\ldots,\bar{r}_n)$ where $\bar{R}$ is the He^3 coordinate and $\bar{r}_1,\ldots,\bar{r}_n$ are the He^4 coordinates. The wave function for an excitation of momentum $\hbar\bar{k}$ in the liquid is taken in the form $F\Phi$ as suggested by Feynman[5] and Cohen.[6] We choose

$$F = a\exp(i\bar{k}\cdot\bar{r}) + b\Sigma_j f(\bar{R}-\bar{r}_j)\exp(i\bar{k}\cdot\bar{r}_j) \qquad (1)$$

where the ratio a/b is to be determined by minimizing the energy and f(r) is a smooth function which vanishes for r greater than some distance d. The effects of backflow have been neglected.

With this choice of $F\Phi$ we obtain the energy of the excitation (neglecting derivatives of f) as

$$E_k = \frac{\hbar^2K^2}{2m_3} \quad \frac{a^2 + \alpha^{-1}b^2I_1}{a^2 + 2abI_2 + b^3(I_1 + I_3)} , \qquad (2)$$

with

$$I_1 = \int d^3r\, P_{34}(\bar{r}) f^2(r), \quad I_2 = \int d^3r\, P_{34}(\bar{r}) f(r) \exp(i\bar{k}\cdot\bar{r}) ,$$

$$I_3 = \int d^3r_1 d^3r_2\, P_{344}(\bar{r}_1,\bar{r}_2) f(r_1) f(r_2) \exp[i\bar{k}\cdot\bar{r}_{12}] . \qquad (3)$$

In (3); $P_{34}(\bar{r})$ is the probability of finding a He^4 at $\bar{r}$ and $P_{344}(\bar{r}_1,\bar{r}_2)$ is the probability of finding a He^4 at $\bar{r}_1$ and another at $\bar{r}_2$ if in both cases there is a He^3 atom at the origin; α is the mass ratio, $\alpha = m_4/m_3 = 1.33$. Minimizing (2) with respect to a/b and choosing the lower energy state (a/b > o) yields

$$E_k = \frac{\hbar^2 k^2}{2m_3} \frac{1}{1 + G_k} , \quad G_k = \frac{2\alpha I_2^2}{(D^2 + 4\alpha I_1 I_2^2)^{1/2}} - D , \qquad (4)$$

where $D = \alpha(I_1+I_3) - I_1$. We may approximate the integrals in (3) as follows: Choose $f(r) = 1$ for $r < d$ and zero otherwise. When $d > r_o$, the interparticle spacing, I_1 becomes a measure of the number of He^4 atoms in the excitation. This is left as a variable parameter. For $kd > 1$, a good approximation is $I_2 \simeq P_{34}(k)$. For I_3 we use the superposition approximation $n_o P_{344}(\bar{r}_1 \bar{r}_2) \simeq P_{34}(\bar{r}_1) \cdot P_{34}(\bar{r}_2) P_{44}(\bar{r}_{12})$ where n_o is the He^4 number density; for $kd > 1$, we have $I_3 \simeq I_1 P_{44}(k)$. The energy is not very sensitive to I_1 for $I \leq I_1 \leq 8$. For I_1 very large, $G_k = \alpha[1 + P_{44}(k)]-1$ and E_k goes over into the roton energy found earlier by Feynman.[5] For further details the reader is referred to our original paper.[7]

The results are: The He^3 spectrum for k near the minimum is represented by $E_k = \Delta_3 + (\hbar^2/2\mu_3)(k-k_3)^2$, with $k_3 = 1.9$ Å^{-1}. We find $G_{k3} = 0.64$, $\Delta_3 = 16.5$ K and a/b = .4 for $I_1 = 2$. The roton gap Δ_4 calculated omitting backflow is 19K and the difference $\Delta_4 - \Delta_3 \simeq 2.5$K. The important result is that the He^3 excitation curve lies below the roton spectrum. A sketch of the proposed spectrum is given in Fig. 1. The effective mass $\mu_3 > \mu_4$ corresponds to the fact that the calculated $P_{34}(k)$ is broader than $P_{44}(k)$, although there is reason to believe that this is untrue.

To summarize: The energy spectrum of He^3 excitations has the form $E_k = \hbar^2 k^2/2m_3^*$ for small wave vectors and $E_k = \Delta_3 + \hbar^2(k-k_3)^2/2\mu_3$ for $k \simeq 1.9$ Å^{-1}. For convenience we refer to these latter excitations as He^3-rotons. As rotons are localized, normal rotons may also exist in mixtures up to concentrations of 31%.

Following Pitaevsky's suggestion, Sobolev and Esel'son[8] reinterpreted their original data[3] and found that for the He^3 quasiparticle spectrum; $\Delta_3 = 6.6$°K, $k_3 = 1.95$ Å^{-1}, $\mu_3 = .07\, m_3$, and $m_3^* = 2.3 m_3$, which is to be compared with the pure He^4 roton parameters; $\Delta_4 = 8.8$K, $k_4 = 2.0$Å^{-1}, $\mu_4 = 0.16 m_4$. What we have named a He^3-roton,

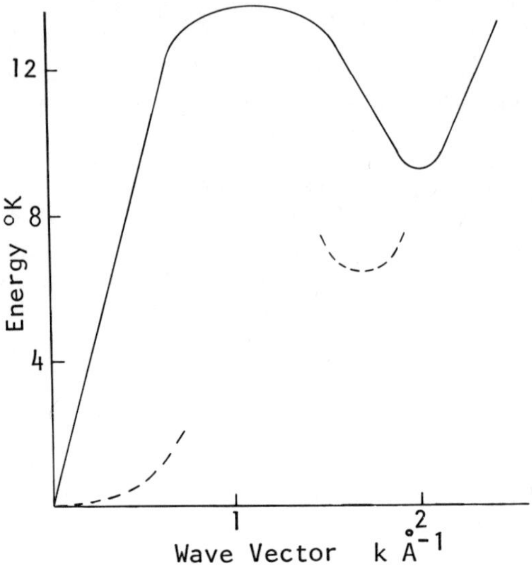

Figure 1

Sobolev and Esel'son[8] call an "i-roton," "i" standing for impurity. They consider the difference $\Delta_4 - \Delta_3 \simeq 2K$ to be a binding energy. This interpretation leads to an attraction between rotons and low energy He^3 -quasiparticles producing a He^3 -roton; the excess energy and momentum being carried away by a phonon, i.e.,

$$He^3\text{-roton} + \text{phonon} \rightleftarrows \text{roton} + He^3\text{-quasiparticle},$$

the quasiparticle having low energy and momentum. With the present experimental accuracy it is difficult to determine whether this reaction goes.

In treating their data, Sobolev and Esel'son[3] found that the He^3 effective mass m_3^* was a function of concentration and satisfied the following empirical formula

$$m_3^*(x) = m_3^*(0)[1+(m_3^*(0)/m_3-1)x]^{-1} \quad , \tag{5}$$

with x the He^3 concentration by weight. Without this refinement, we find the He^3 impurity contribution to the normal fluid density ρ_{ni}, the energy density E_i, and the chemical potential $\overline{\mu}_3$ are given by (in the non-degenerate case)

$$\rho_{ni} = n_3 m_3^* \left[\frac{1 + \frac{1}{3}\left(\frac{\hbar^2 k_3^2}{m_3^* kT}\right) f}{1 + f} \right] ,$$

$$\beta E_i = \frac{3}{2} \left[\frac{1 + \frac{2}{3}(\beta\Delta_3 + {}^1/_2) f}{1 + f} \right] ,$$

$$\beta m_3 \bar{\mu}_3 = \ell n(n_3) - \frac{3}{2} \ell n\left(\frac{m_3^* kT}{2\pi\hbar^2}\right) - \ell n(1+f) \quad , \qquad (6)$$

where $\beta = (kT)^{-1}$ and $f = (2\hbar^2 k_3^2/m_3^* kT)(\mu_3/m_3^*)^{1/2} \exp(-\beta\Delta_3)$. The He^3 chemical potential can be obtained from a measurement of the He^3 concentration in vapor in equilibrium with the liquid. For the He^3 dynamic structure factor $S_3(k,w)$ in both the degenerate and non-degenerate cases the reader is referred to Ref. 7.

The possibility exists of observing a Raman-scattering process in which two He^3-rotons are created. The ratio of the intensities of the two-He^3-roton and the two-roton Raman spectra is approximately $(n_3/n_4)^2$ where n_3 and n_4 are the He^3 and He^4 densities.

Raman scattering in which a He^3-roton near k_3 and a He^4-roton near k_4 (where the densities of states are largest) is forbidden by momentum conservation when $k_4 > k_3$. Raman scattering at a frequency shift of $\omega \simeq \Delta_3 + \Delta_4$, would show no special features. It has been suggested that advances in neutron scattering techniques may yet demonstrate the He^3 energy spectrum directly.

References

(1) C. M. Surko and R. E. Slusher, Phys. Rev. Lett. 30, 1111 (1973).

(2) R. L. Woerner, D. A. Rockwell, and T. J. Greytak, Phys. Rev. Lett. 30, 1114 (1973).

(3) V. I. Sobolev and B. N. Esel'son, Zh. Eksp. Teor. Fiz. 60, 240 (1971) [Sov. Phys. JETP 33, 132 (1971)].

(4) L. P. Pitaevsky, in proceedings of the U.S. - Soviet Symposium on Condensed Matter, Berkeley, California, May, 1973.

(5) R. P. Feynman, Phys. Rev. 94, 262 (1954).

(6) R. P. Feynman and M. Cohen, Phys. Rev. 102, 1189 (1956).

(7) M. J. Stephen and L. Mittag, Phys. Rev. Lett. 31, 923 (1973).

(8) V. I. Sobolev and B. N. Esel'son, ZhETF Pis. Red. 18, 689 (1973) [JETP Lett. 18, 403 (1973)].

RIPPLONS AND SUPERFLUIDITY IN ^{3}He MONOLAYERS*

Chia-Wei Woo

Department of Physics, Northwestern University

Evanston, Illinois 60201

Abstract: In our continuing effort to establish the existence of superfluidity in He^3 monolayers physisorbed on liquid He^4, we calculated the ripplon spectrum in liquid He^4, and obtained wave functions describing ripplons. The latter, when exchanged between He^3 quasiparticles, are expected to enhance the pairing mechanism and thereby raise the transition temperature.

In an effort to establish the existence of superfluidity in He^3 monolayers physically adsorbed on liquid He^4, we embarked on a five-part theoretical program approximately a year and a half ago. In the 30 minutes allocated to me here, I wish to describe briefly the third part of that program.

First of all, the crudest model that one can construct consists of a layer of He^3 atoms moving horizontally on a sharply defined, static surface of liquid He^4. Thus we deal with a two-dimensional He^3 liquid. Our earlier work[1] indicated a transition temperature somewhat lower than that for bulk He^3. Such a model is actually more suitable for describing He^3 physisorbed on grafoil. It is totally unsatisfactory for the system under consideration. In the present study, then, we refine the model by asking the following questions in sequence:

1. What does the He^4 surface really look like?
2. Will a He^3 atom stick (float) on such a surface?
3. Is there anything happening in the surface layer to affect

*Work supported in part by the National Science Foundation through grants GH-39127 and DMR74-09661.

the behavior of the He^3 atom?
4. How does a He^3 monolayer behave?
5. What is the superfluid transition temperature?

To answer the first question, one calculates the density profile $\rho(z)$ of liquid He^4 near its free surface. A crucial quantity here is the surface tension σ, which has an experimental value of 0.378 dyn/cm at zero temperature. Our variational calculation[2] yielded a value of 0.360 dyn/cm. Similar calculations by Chang and Cohen[3,4] yielded 0.414 dyn/cm. Padmore and Cole[5] obtained $\rho(z)$ by way of a density functional approach while fitting σ to the experimental value, and most recently Ebner and Saam[6] improved upon the latter, obtaining a surface tension of 0.384 dyn/cm. All these calculations made use of local approximation schemes for the pair correlation function $g(\vec{r}_1,\vec{r}_2)$, to which we shall return later. We feel that the problem has by now been well solved. There still remains some concern as to whether $\rho(z)$ contains an overshoot and oscillations. I tend to believe the conclusion of Ebner and Saam that while there are oscillations near a liquid-solid interface no oscillations exist at the free surface. We are doing more work ourselves to confirm it.

To answer the second question, one looks for surface states of He^3 in He^4. The crucial quantity there is the excess binding energy of He^3 in the surface layer. Saam[7] obtained 1.26K, we[2] 1.6K, and Chang and Cohen[4] 2.45K, in comparison to the experimental value of 1.7-1.95K. We consider that problem also essentially solved.

My talk at the Technion[1] conference last summer covered the two topics above.

I plan to address to the third question here. We are interested in studying the excitations in He^4: phonons and ripplons, and how they dress a physisorbed He^3 atom. The major contributor to the work reported here is A. Bhattacharyya. Part of her Ph.D. research is on this problem. A second collaborator is Y. M. Shih.

Question No. 4 deals with the effective interaction between a pair of He^3 atoms, in particular how the exchange of phonons and ripplons makes the interaction more attractive. Once the interaction is known, it is possible to calculate the normal-liquid properties and compare them to experiment. One also learns the maximum monolayer coverage, as well as the maximum areal density at which the monolayer remains fluidic.

One encouraging feature of our theoretical program is that up to this point every step can be checked against experiment before proceeding to the next. Only after we become thoroughly convinced of the accuracy of the preparatory work will we tackle the final task of predicting the superfluid transition temperature. I hope to be able to report on Parts 4 and 5 at a later date, say at LT14 this summer.

The theory of classical capillary waves can be summarized in a few sentences. Combining the equations of motion, continuity, and state for potential flow, one obtains a wave equation governing the velocity potential $\Phi(\vec{r},t)$. For a liquid with a surface, however, it is not easy to solve the wave equation under proper boundary conditions, so instead one employs Laplace's formula at the free surface. Guessing the solution in the form

$$\Phi(\vec{r},t) = Ae^{\kappa z} e^{i(\vec{k}\cdot\vec{\rho}-\omega t)} , \tag{1}$$

where $\vec{k}$ and $\vec{\rho}$ are two-dimensional vectors, and substituting into the wave equation (which must be satisfied deep in the bulk) and Laplace's formula (which must be satisfied near the surface), one obtains the relations:

$$k^2 - \kappa^2 = \frac{\omega^2}{c^2} , \tag{2}$$

and

$$\omega^2 = \frac{\sigma}{m\rho_o} k^2\kappa . \tag{3}$$

At long wavelengths, i.e., $k \to 0$, the speed of sound c is constant. One eliminates ω^2 and finds $\kappa \to k$, and thus $\omega \propto k^{3/2}$: the well-known dispersion law for capillary waves. For details one can look up Ref. 8.

A semi-quantum-mechanical version of the above can be constructed from Gross' Hartree liquid theory.[9] Gross describes the liquid by a Hartree function:

$$\Psi(1,2,\ldots,N;t) \approx \prod_{i=1}^{N} \frac{1}{\sqrt{N}} \psi(\vec{r}_i,t) . \tag{4}$$

The normalization of Ψ results in $|\psi|^2$ being interpreted as the local density $\rho(\vec{r},t)$. Writing

$$\psi(\vec{r},t) = R(\vec{r},t)\exp[\frac{im}{\hbar} \chi (\vec{r},t)] \tag{5}$$

and substituting it into the self-consistent equation

$$i\hbar \frac{\partial\psi(\vec{r},t)}{\partial t} = \frac{-\hbar^2}{2m} \nabla^2\psi(\vec{r},t) + \psi(\vec{r},t)\int v(\vec{r}-\vec{r}\,')|\psi(\vec{r}\,',t)|^2 d\vec{r}\,', \tag{6}$$

the latter separates into

$$\frac{\partial R^2}{\partial t} = -\nabla[R^2\nabla\chi] \ , \tag{7}$$

and

$$-\frac{\partial\chi}{\partial t} = \frac{1}{2}(\nabla\chi)^2 + \pi, \tag{8}$$

where

$$\pi(\vec{r},t) = \frac{-\hbar^2}{2m^2}\frac{\nabla^2 R}{R} + \frac{1}{m}\int R^2(\vec{r}',t)v(\vec{r}-\vec{r}')d\vec{r}' \ . \tag{9}$$

The identification of R^2 with $\rho(\vec{r},t)$ implies that χ corresponds to the velocity potential Φ. Equations (7) and (8) are then the equations of continuity and motion.

Let

$$\rho(\vec{r},t) = \underline{\rho_o(\vec{r})} + \tilde{\rho}(\vec{r},t) \ , \tag{10}$$

and

$$\Phi(r,t) = \underline{\Phi_o(\vec{r}) - \frac{E}{m}t} + \tilde{\Phi}(\vec{r},t) \ , \tag{11}$$

where the underlined portions represent the stationary solution. Substituting Eqs. (10) and (11) into Eqs. (7) and (8), linearized equations are derived for $\tilde{\rho}$ and $\tilde{\Phi}$. Solving the latter for a weakly interacting Bose gas as an illustration, one finds

$$\tilde{\Phi}(\vec{r},t) = Ae^{i(\vec{q}\cdot\vec{r}-\omega t)} \tag{12}$$

with

$$\hbar\omega = \{(\frac{\hbar^2 q^2}{2m})^2 + \frac{\rho_o\hbar^2 q^2}{m} v(q)\}^{1/2}, \tag{13}$$

the Bogoliubov spectrum.

For application to a liquid with a free surface, one takes

$$\tilde{\Phi}(\vec{r},t) = Ae^{\kappa z}e^{i(\vec{k}\cdot\vec{\rho}-\omega t)} \tag{14}$$

and finds that deep in the interior

$$\hbar\omega = \{[\frac{\hbar^2(k^2-\kappa^2)}{2m}]^2 + \frac{\rho_o\hbar^2(k^2-\kappa^2)}{m} v(\sqrt{k^2-\kappa^2})\}^{1/2}, \tag{15}$$

as expected. At the surface, if once again Laplace's formula is employed, one has

$$\omega^2 = \frac{\sigma}{m\rho_o} k^2\kappa \quad . \tag{16}$$

Together Eqs. (15) and (16) yield $\kappa(k)$, and the excitation spectrum $\omega(k)$.

Edwards, Eckhardt, and Gasparini[10] used the above approach in the analysis of their surface tension data. By using the empirical phonon-roton spectrum instead of the Bogoliubov formula in Eq. (15), they found a ripplon spectrum which starts out as $k^{3/2}$, reaches a maximum, and then falls into the roton dip.

We[11] have constructed a fully quantum-mechanical theory. First, take the excited state wave function in the form

$$\psi = \psi_o \, e^{i \sum\limits_{\ell=1} \Phi(\vec{r}_\ell)} \quad , \tag{17}$$

where ψ_o is the ground state and $\Phi(\vec{r})$ is the spatial part of the velocity potential. This is customary. It can be justified by calculating the current and comparing it to density times $\nabla\Phi$. Expanding the exponential, retaining the term linear in Φ, and realizing that translational symmetry exists in the horizontal plane, one writes after projecting out ψ_o:

$$\psi \approx \sum_{\ell=1}^{N} f(z_\ell) e^{i\vec{k}\cdot\vec{\rho}_\ell} \psi_o \quad . \tag{18}$$

This wave function was suggested by Miller and myself[12] and used also in a recent paper by Chang and Cohen.[13] From the expectation value of H in ψ one obtains the excitation spectrum

$$\omega(k) \approx \frac{\langle\psi|H|\psi\rangle}{\langle\psi|\psi\rangle} - \frac{\langle\psi_o|H|\psi_o\rangle}{\langle\psi_o|\psi_o\rangle} \quad . \tag{19}$$

The Euler-Lagrange equation[12,13] which minimizes $\omega(k)$:

$$\frac{h^2}{2m}\left[f''(z)+\frac{\rho'(z)}{\rho(z)} f'(z)-k^2 f(z)\right]+\omega(k)\left[f(z) + \frac{1}{\rho_o}\int_{-\infty}^{\infty} f(z')\rho(z')g_k(z,z')dz'\right] = 0 \quad , \tag{20}$$

where ρ_o is the bulk equilibrium density and $g_k(z,z')$ is the Fourier transform of $g(\vec{r}_1,\vec{r}_2)$ weighted by the density profile, takes the place of Gross' self-consistent equation. The problem can be solved approximately by setting f(z) to $e^{\kappa z}$. Deep in the interior, Equation (20) yields the Feynman phonon spectrum

$$\hbar\omega(k) = \frac{\hbar^2(k^2-\kappa^2)}{2mS(\sqrt{k^2-\kappa^2})} , \tag{21}$$

where S is the liquid structure function. At the surface, Equation (19) gives

$$\hbar\omega(k) = \frac{\hbar^2(k^2+\kappa^2)}{2m} \{1 + \frac{\int\int_{-\infty}^{\infty} e^{\kappa(z+z')}\rho(z)\rho(z')g_k(z,z')dzdz'}{\rho_0\int_{-\infty}^{\infty} e^{2\kappa z}\rho(z)dz}\}^{-1}$$

Simultaneous solution of Eqs. (21) and (22) leads to $\kappa(k)$ and $\omega(k)$. It turns out that there exist two solutions. One of which, $\kappa = 0$, yields the phonon-roton spectrum. The other one starts out as $k^{3/2}$, rises to a peak, and falls into the roton dip, much like the spectrum predicted by Edwards et al.[10] It is the ripplon. We expect that when ripplon-ripplon and ripplon-phonon scattering processes are taken into consideration in a perturbation calculation, the spectrum will be found in good quantitative agreement with the semi-empirical curve of Ref. 10.

Preliminary calculations using the unrenormalized phonon and ripplon energies and wave functions lead to an effective mass of 1.34 m_3 for a physisorbed He^3 quasiparticle. (The ripplon alone accounts for 1.07 m_3.) The experimental value from Edward's group is 1.3 ± 0.1 m_3. It is not quite as big as I'd like, since a large effective mass suggests strong He^3-ripplon and He^3-phonon vertices, and consequently a strong substrate-mediated attraction.

References

(1) C.-W. Woo, *Proceedings of the European Physical Society Topical Conference: Liquid and Solid Helium*, Technion, Haifa, Israel (1974). The collaborators on the work referred to are A. Bhattacharyya and M. Schick.

(2) Y. M. Shih and C.-W. Woo, Phys. Rev. Lett. 30, 478 (1973).

(3) C. C. Chang and M. Cohen, Phys. Rev. A8, 1930 (1973).

(4) C. C. Chang and M. Cohen, Phys. Rev. A8, 3131 (1973).

(5) T. C. Padmore and M. W. Cole, Phys. Rev. A9, 802 (1974).

(6) W. F. Saam and C. Ebner, Phys. Rev. Lett. 34, 253 (1975).

(7) W. F. Saam, Phys. Rev. A4, 1278 (1971).

(8) L. D. Landau and E. M. Lifshitz, Fluid Mechanics, Pergamon Press (1959), Chap. VII.

(9) E. P. Gross, J. Math. Phys. 4, 195 (1963).

(10) D. O. Edwards, J. R. Eckhardt, and F. M. Gasparini, Phys. Rev. A9, 2070 (1974).

(11) A. Bhattacharyya, Y. M. Shih, and C.-W. Woo, submitted to Phys. Rev. Lett.

(12) M. D. Miller, Ph.D. Thesis, Northwestern University (1973). See also H. W. Lai, C.-W. Woo, and F. Y. Wu, J. Low Temp. Phys. 3, 463 (1970).

(13) C. C. Chang and M. Cohen, Phys. Rev. B, to be published.

FILM THICKNESS DETERMINATIONS IN MOVING SATURATED SUPERFLUID ^{4}He FILMS*

Robert B. Hallock†

Department of Physics, Brown University,
Providence, Rhode Island 02912

and

Department of Physics and Astronomy,
University of Massachusetts
Amherst, Massachusetts 01002‡

Abstract: Preliminary results of precision measurements of the velocity dependent saturated film thickness are presented. We find agreement within experimental errors with predictions based on the theory of Kontorovich. These measurements rule out a strong velocity dependence for ρ_s/ρ in saturated films. We also report the first observations of thickness changes due to third sound in saturated films by use of direct capacitance techniques. At 1.46K the attenuation of third sound at 1.2 Hz is found to be about $0.008\ cm^{-1}$.

Kontorovich[1] was the first to point out that a saturated film of superfluid ^{4}He in uniform flow under isothermal conditions should thin with increasing velocity. Keller[2] reported the first direct capacitance measurements of the film thickness. No thinning was observed as a function of film velocity in his experiments and this fact has stimulated both theoretical[3-6] and experimental[7-15] work.

---

*Supported at the University of Massachusetts by the National Science Foundation (GH34534A1).

†A. P. Sloan Fellow.

‡Permanent address.

The majority of recent experiments report the observation of thinning in general agreement (usually within 50%) with the Kontorovich[1] prediction. The present experimental study has been undertaken in an attempt to measure the thinning to high enough accuracy to allow a quantitative test of the Kontorovich[1] prediction. To the extent the results have been analyzed, they do show such quantitative agreement with a sensitivity of ± 0.6 Å. During the course of this study it became clear that third sound[16] resonances could be observed in the apparatus by direct capacitance techniques. We report here the first such observations. Precision direct capacitance methods should allow detailed studies of third sound in thick helium films--a region which has until now been accessible only with great difficulty[17] and hence almost completely neglected.[18]

The apparatus, Fig. 1, is patterned to have the general characteristics of the apparatus used by Keller[2,10]. In the present case

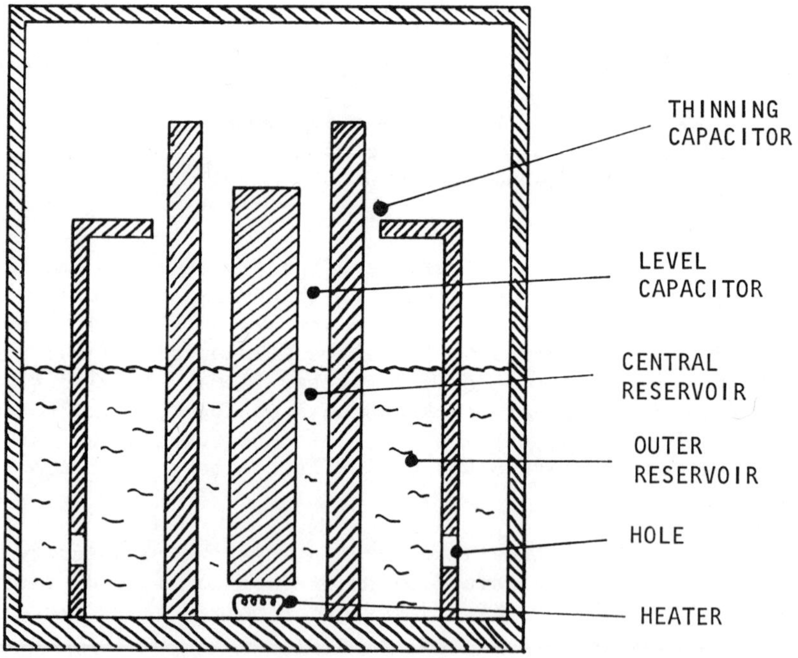

Figure 1. Schematic of the thinning capacitor and level detector. Oscillations and level changes between the two reservoirs are initiated by means of the heater in the central reservoir. A heater (not shown) in the outer reservoir can also be used. The chamber is sealed by means of a superfluid valve. The chamber is immersed in He II inside a second chamber located at the end of a standard dewar insert. Level changes can be resolved to 200Å and film thickness changes can be resolved to 0.6Å.

construction is entirely of OFHC[19] copper. The general idea is that a change in helium level difference between the two reservoirs will result in an approach to equilibrium by means of film flow and an oscillation in reservoir levels about the equilibrium level difference.[20] Direct capacitance measurements of the film thickness[2,8] can then be made both with the film in motion and at rest. The present design utilizes a thinning capacitor fashioned from a single piece of metal. This is to avoid possible multiple flow paths on the electrode which could allow persistent currents[21,22,23]. This design therefore necessitates corrections to the observed capacitance due to the small but non-negligible parallel capacitance between the sides of the thinning capacitor and the central reservoir. In what follows explicit mention of these corrections will not be made although they have been applied.

The calibration of the velocity of film flow is carried out by use of the level capacitor. Given the film thickness in the region of the thickness capacitor one simply uses conservation of mass and the known volume of the level capacitor per unit length to obtain the velocity of the film v_s. The level capacitor is carefully machined[24] and of known[25] (I.D. = 0.635 cm) dimensions. This capacitor also has calibration marks to allow direct level vs capacitance determinations. The level capacitor sensitivity was 0.926 pf/cm. The thickness capacitor was calibrated by observing the capacitance change which resulted from the addition of helium gas to the experimental chamber. This observation was carried out at several temperatures and no temperature dependence was seen. Our procedures were patterned after and were entirely consistent with those followed by Keller.[2] For the present apparatus the sensitivity of the film thinning capacitor gap to atoms in the gap was found to be 1.61×10^{22} atoms/cm^3 pf. As in Keller's[2] experiments, a marked increase in capacitance signaled the formation of the saturated film. Further addition of small amounts of helium gas caused no observable change in capacitance once equilibrium was reached.

The helium gas used for the calibrations and used to fill the reservoirs with bulk liquid was passed through a Linde[26] 13X molecular sieve trap at liquid nitrogen temperature and through a millipore[27] filter at the operating temperature of the cryostat ($T < T_\lambda$). The metal surfaces on which the thinning measurements were made were carefully lathe turned and then mechanically polished with crocus cloth. No attempt has yet been made to electropolish the flow surfaces.[28]

Oscillations are induced by the application of heat (0.1 - 2.0 mw) to the central reservoir. The resulting fountain pressure causes a flow to the central reservoir and an increase in its level. Oscillations about the fountain induced equilibrium level rise take place. The removal of the heat causes a reverse flow and again oscillations.

This can be seen by reference to Fig. 2. Also shown are the observed thickness changes of the film in the thinning capacitor.[29] The thinning is all assumed to take place on the wall of the central reservoir since the film on the thinning electrode is assumed not to move. An asymmetry is observed[30] in the heater-on traces and is due to the presence of a film flow up the outer wall of the central reservoir. This flow is necessary to balance the atoms lost from the central reservoir due to evaporation. Typically temperature differences between the reservoirs of a few tens of μK result from the application of heat. The asymmetries caused by the bias film flow when the heater is on are similar to asymmetries observed by Williams and Packard[8] and we have pointed out previously[29] that persistent currents are a likely source for their observations.

An example of a heater-off thinning trace which has been analyzed in detail is shown in Fig. 3. Given the present uncertainties in the thinning expected for our apparatus (due to the design of the capacitor as previously described), what is actually observed is in agreement with the Kontorovich[1] prediction. The errors quoted on the figure are conservative since the analysis is still preliminary. The effect on the thinning of acceleration terms[9] can be shown to be negligible in the present apparatus. Since the present results are in quantitative agreement with expectations based on the Kontorovich[1] prediction, it is clear that the velocity dependence assigned[13] to ρ_s/ρ to explain the differences between the observations of Flint and Hallock[9] on one hand and Telschow, Wang and Rudnick[13] on the other is not observed in the present experiments. Our observations to the extent they have been analyzed are entirely consistent with the thinning predicted by

$$\delta = \delta_o \left(1 + \frac{\rho_s v_s}{2\rho g z}\right)^{-1/3} , \tag{1}$$

with ρ_s/ρ independent of velocity. Here δ is the film thickness at a height z above the free surface when the film has velocity v_s. g is the acceleration due to gravity. The measurements shown in Fig. 3 were made at a height z = 1.1 cm between the free surface and the center of the thinning gap. At this level our observations are consistent with the expression $\delta_o = 3.4 \times 10^{-6} z^{-1/3}$ where δ_o is the static thickness in centimeters. Our films thus appear to be somewhat thinner than those observed by Keller.[2] Precise determinations of the exponent cannot be made in the present apparatus.

An example of a third sound resonance is shown in Fig. 4. Plotted is the capacitance seen in the thinning gap after a heat pulse to a heater located on the outer wall of the thinning capacitor. The period of the oscillation at 1.46K is 0.86 sec and the

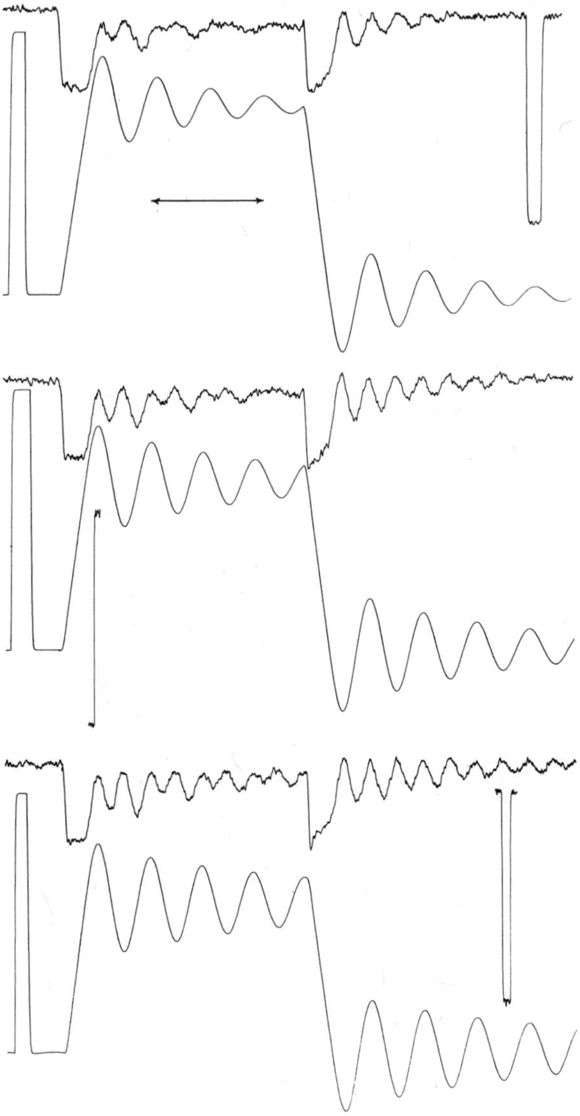

Figure 2. Examples of level changes in the central reservoir (smooth traces) accompanied by changes in the film thickness at the position of the thickness capacitor. Observed capacitance increases vertically in both cases. The calibration offsets correspond to a level change of 2.16 mm and a thickness change of 35Å. The horizontal bar represents 60 sec. In the three examples T = 1.602K (top), 1.500K and 1.419K a power of 0.47 mw is first applied (left hand portion of figure) and then removed from the heater in the central reservoir.

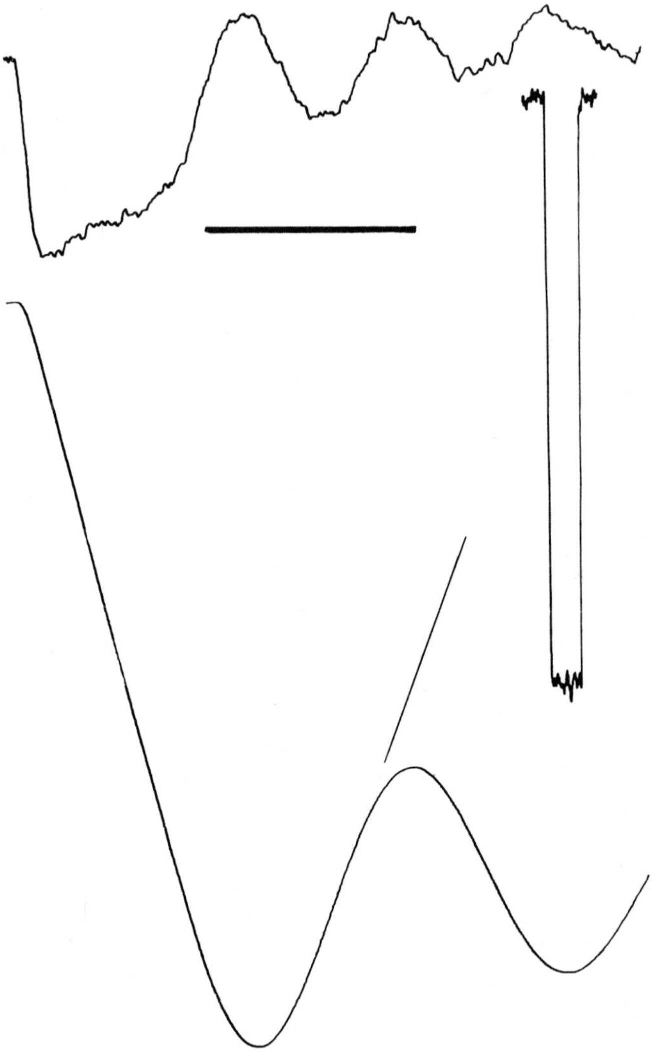

Figure 3. Film thinning observed in response to the shut off of the heater at 1.602K. The horizontal bar represents 20 sec. The capacitance calibrations are the same as in Fig. 2. The slope of the level change curve indicated corresponds to a film velocity of 12.2 cm/sec in the region of the gap of the thinning capacitor. An average of a number of such measurements results in an observed thinning of 6.07 ± 0.64Å for the velocity of 12.2 cm/sec. The Kontorovich expression, eq. 1, predicts a thinning of 6.2 ± 0.6Å for the conditions of the measurement. The uncertainty in the prediction is discussed in the text.

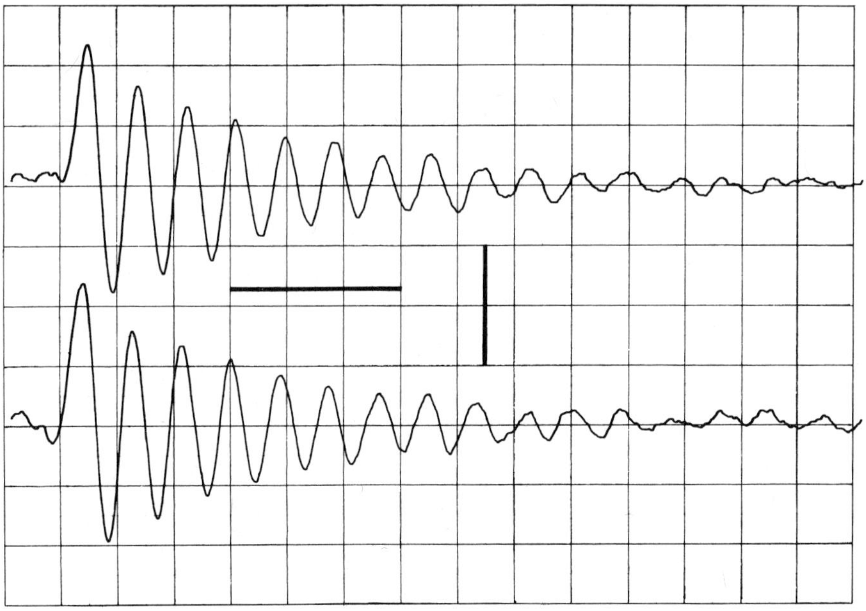

Figure 4. Examples of thickness oscillations observed in the thinning capacitor at T = 1.461K. We attribute this observation to the presence of third sound. The oscillations are initiated in this example by a heater. The horizontal bar represents 3 sec; the vertical bar represents a thickness change of 13Å.

attentuation, α, is calculated to be $\alpha \simeq 0.008$ cm^{-1}. Note that since the apparatus was not designed for such studies, the film path was not of uniform film thickness and an average velocity must be selected to obtain the above attenuation from the observed decay of the oscillation. We expect therefore that there may be as much as a factor of two error associated with α. The point we wish to make here however is not the specific value of α, but rather the fact that using direct capacitance techniques one can study third sound in relatively thick saturated films. We expect to use this technique and others like it in the future to explore third sound in the much neglected domain of thick films.

Still open and not answered by this work are the questions of why third sound techniques[13] show no thinning as a function of velocity and why no thinning was observed in the original experiment of Keller.[2]

References

(1) V. M. Kontorovich, Zh. Eksp. Theor. Fiz. 30, 805 (1956). [Soviet Phys. - JETP 3, 770 (1956)].

(2) W. E. Keller, Phys. Rev. Lett. 24, 569 (1970).

(3) E. R. Huggins, Phys. Rev. Lett. 24, 573 (1970).

(4) D. L. Goodstein and P. G. Saffman, Proc. R. Soc. Lond. A 325. 447 (1971).

(5) D. J. Bergman, Proc. R. Soc. Lond. a 333, 261 (1973).

(6) L. J. Campbell, to be published.

(7) E. von Spronsen, H. J. Verbeek, R. de Bruyn Ouboter, K. W. Taconis and H. van Beelan, Phys. Lett. A 45, 49 (1973).

(8) G. A. Williams and R. Packard, Phys. Rev. Lett. 32, 587 (1974).

(9) E. B. Flint and R. B. Hallock, Bull. Am. Phys. Soc. 19, 436 (1974) and Phys. Rev. B (March, 1975).

(10) M. E. Banton, J. K. Hoffer and W. E. Keller, Bull. Am. Phys. Soc. 19, 436 (1974).

(11) G. M. Graham and E. Vittoratos, Phys. Rev. Lett. 33, 1136 (1974).

(12) E. B. Flint, Ph.D. Dissertation, University of Massachusetts, 1974.

(13) K. Telschow, J. Wang and I. Rudnick, J. Low Temp. Phys. (to be published); see also K. Telschow, Ph.D. Dissertation, UCLA, 1973.

(14) E. van Spronsen, H. J. Verbeek, H. van Beelen, R. de Bruyn Ouboter and K. W. Taconis, Physica (to be published).

(15) T. G. Wang and D. Petrac (preprint).

(16) See, for example, K. R. Atkins and I. Rudnick, Progress in Low Temperature Physics Vol. 6, ed. C. J. Gorter, (North Holland, Amsterdam, 1970), Chapter 2, p. 37.

(17) C. W. F. Everitt, K. R. Atkins and A. Denenstein, Phys. Rev. 136, 1494 (1964).

(18) E. van Spronsen, H. J. Verbeek, H. van Beelen, R. de Bruyn Ouboter and K. W. Taconis have reported third sound oscillations in their very long narrow flow capillaries. They study these oscillations by observing bulk level changes in their capacitors.

(19) Oxygen Free High Conductivity.

(20) R. B. Hallock and E. B. Flint, Phys. Rev. A 10, 1285 (1974).

(21) H. J. Verbeek, E. van Spronsen, H. Mars, H. van Beelan, R. de Bruyn Ouboter and K. W. Taconis, Physica (Utrecht) 73, 621 (1974).

(22) R. K. Galkiewicz and R. B. Hallock, Phys. Rev. Lett. 33, 1073 (1974).

(23) L. J. Campbell, Proceedings of Scottish Universities Summer School, (St. Andrews, 1974) in press.

(24) E. B. Flint and R. B. Hallock, Rev. Sci. Instrum. 46, 100 (1975).

(25) Westfield Gauge Co., Westfield, Mass. has the capability to measure inside and outside diameters to an accuracy of about ± 0.2μ.

(26) Union Carbide Corporation.

(27) Millipore Corporation, Bedford, Mass. 01730.

(28) This will be done in the future in an attempt to see if the observation of thinning depends critically on the flow surface in an experiment of this type. Ref. 9 suggests that it will not, but ref. 10 suggests that it may.

(29) The present measurements were made using a General Radio 1615A capacitance bridge and an HR-8 as a null detector. Helium level measurements could be made to a sensitivity of 200Å (although it was not necessary to do so). Relative thinning was measured to a sensitivity of 0.6Å (2×10^{-6} pf).

(30) R. B. Hallock (to be published).

RECENT EXPERIMENTS ON THE SURFACE OF LIQUID ^{4}He:

ELASTIC SCATTERING OF ^{4}He ATOMS*

D. O. Edwards,* P. P. Fatouros, G. G. Ihas,
P. Mrozinski,† S. Y. Shen§ and C. P. Tam

Department of Physics, The Ohio State University

Columbus, Ohio 43210

Abstract: The first part of the paper concerns the experimental determination of the two-dimensional effective interaction potential between ^{3}He atoms adsorbed on the surface of liquid ^{4}He. Measurements of the velocity of surface sound and of the surface tension indicate that, for concentrations of ^{3}He corresponding to a small fraction of a monolayer, the interaction is very weak and predominantly repulsive. This is unfavorable for a low-temperature superfluid condensation.

The bulk of the paper discusses the implications of a recent experiment on the elastic scattering of free ^{4}He atoms at the surface of pure liquid ^{4}He. The elastic scattering probability $R(k,\theta)$ has been measured as a function of the momentum of the atom $\hbar k$ and of the angle of incidence θ. In addition the probability of inelastic scattering was found to be very small (less than 2×10^{-3}).

With the simplifying assumption that the scattered atoms can be treated as distinguishable from those in the liquid it is found that $R(k,\theta)$ can be calculated from the density profile $\rho(z)$ at the surface of the liquid. The relation between the scattering data and the spectrum of atoms evaporated from the liquid at finite temperature is also discussed.

---

*Supported by a grant from the National Science Foundation, GH 31650 A#2, and partially performed under the auspices of the Energy Research and Development Administration.
†Now at duPont Experimental Station, Wilmington, Delaware
§Now at Northwestern University, Evanston, Illinois.

Before starting on the main subject of this paper--the elastic scattering of neutral ^4He atoms from the surface of liquid helium--we would like to append some remarks to C. W. Woo's paper,[1] particularly about the possibility of observing 'two-dimensional' superfluidity in ^3He adsorbed on the surface of liquid ^4He. We have studied the effective interaction between ^3He quasiparticles on the surface of ^4He by making simultaneous measurements of the depression of the surface tension, which is due to the two-dimensional pressure of the ^3He, and the velocity of 'surface sound,' which is a longitudinal ^3He density wave in which the wave motion takes place parallel to the surface.[2] A description of the experimental method and of the theory can be found in Ref. 3, and a more detailed account will soon be published.[4] The measurements were made on two samples with He number densities corresponding to small fractions of a monolayer. We have made a least squares fit between the data and a theoretical model in which the effective interaction is treated in the same way as the Bardeen-Baym-Pines interaction between ^3He quasiparticles in three-dimensions.[5,6,7] The interaction is represented by the Fourier transform $V^s(k)$ which is expanded in even powers of k: $V^s(k) = V^s_0(1+\alpha k^2+...)$. Owing to the weakness of the interaction (the system is almost an ideal 2D Fermi gas) we are only able to determine the first term in this expansion, V^s_0. The results of the fit are shown in Fig. 1. This shows the "χ^2 probability," which is the probability that the deviations from the theory are due purely to random experimental errors, as a function

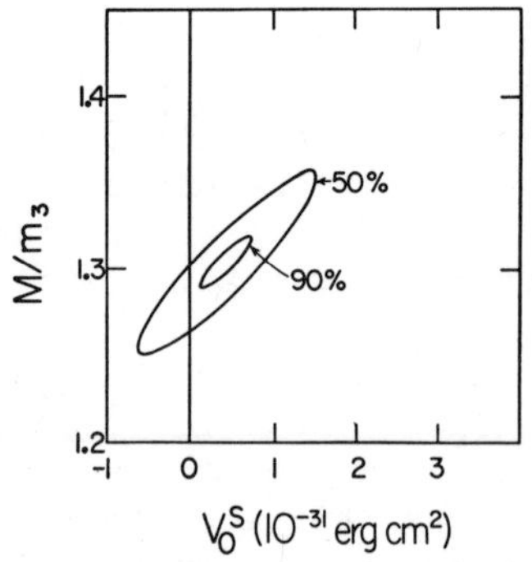

Figure 1. Least-squares fit of the interacting 2D gas model to data on the velocity of surface sound and on the surface tension in two samples of ^3He adsorbed on ^4He. The graph shows contours of the χ^2 probability as a function of the quasiparticle effective mass M and the interaction parameter V^s_0 (see text). A positive value of V^s_0 corresponds to an interaction which is predominantly repulsive at large distances. (From Ref. 4).

of the assumed values of M, the effective mass of a single surface quasiparticle, and V_o^s, the interaction parameter. (A high χ^2 probability corresponds to a good fit to the data.) We see that V_o^s is very small and probably positive. A positive V_o^s corresponds to a repulsive interaction which is unfavorable for superfluidity. The smallness of the interaction can be illustrated in the following way: In 3D it is customary to compare V_o with the quantity $m_4 s^2/n_4$ where s and n_4 are the velocity of sound and number density in pure ^{4}He. At the saturated vapor pressure one finds $V_o \simeq 0.08 m_4 s^2/n_4$. In comparison, the corresponding quantity for the surface $0.08 m_4 s^2/n_4^{2/3} = 8 \times 10^{-31}$ erg cm^2, an order of magnitude larger than the most probable value in Fig. 1.

The fact that V_o^s is probably positive is not, of course, conclusive, as Woo has pointed out, since higher order terms in the expansion of $V^s(k)$ could be attractive and could lead to a superfluid condensation, perhaps in pair states with $\ell \neq 0$. At present we are trying to make a systematic study of the surface tension and surface sound velocity over a much wider range of surface ^{3}He density so as to narrow down some of these possibilities. Unfortunately, because of damping by the ^{3}He dissolved in the bulk of the liquid, surface sound cannot be propagated in densities over ~1 monolayer. Hence the interesting region[8] where the thickness of the sample approaches the coherence length in bulk superfluid ^{3}He (a few hundred angstroms) can only be investigated by measuring the surface tension. There are also formidable difficulties in cooling ^{3}He-^{4}He solutions to the temperature of the transition in pure ^{3}He, since the Kapitza boundary resistance is much larger[9] than in pure ^{3}He.

We now describe the results of an experiment that we have recently completed[10] to study the surface of pure ^{4}He: the measurement of the probability of elastic scattering $R(k,\theta)$ for ^{4}He atoms striking the surface of the liquid as a function of their momentum $\hbar k$ and angle of incidence θ. During the measurements the temperature of the liquid was varied between 0.025 K and 0.125 K and no temperature dependence was found, so that the values of $R(k,\theta)$ are characteristic of the ground state of the liquid and its surface. The experimental method and results are described in Ref. 10 and an account of some preliminary experiments is given in Ref. 11. The present paper is confined to a discussion of the results and their theoretical interpretation. More complete references to previous work can be found in Ref. 10.

When we began the experiment some years ago[11] there were some definite theoretical predictions[12-16] to test, predictions which had been made in connection with the theory of evaporation from liquid helium. The way in which our experiment is related to evaporation is illustrated by Fig. 2. There are three possibilities for an atom

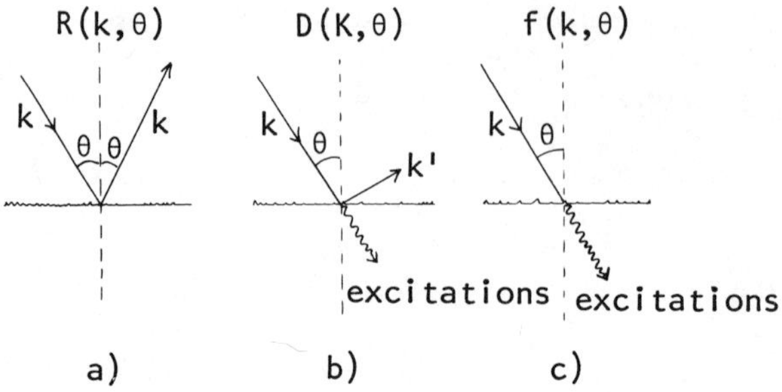

Figure 2. The three possible fates for a free ^{4}He atom striking the surface of the liquid and their probabilities: (a) elastic scattering (specular reflection), (b) inelastic scattering, converting part of the atom's kinetic energy into excitations (phonons and ripplons) in the liquid, (c) absorption, creating excitations whose total energy is equal to the kinetic energy of the atom plus the binding energy, L_o .

striking the surface: (a) Specular reflection (elastic scattering) in which no energy is transferred to the liquid. We define the probability of this occurring as R(k,θ). (b) Inelastic scattering in which the atom rebounds back into the vacuum above the liquid but with loss of energy to the liquid. We call the probability of this process D(k,θ). (c) Absorption or condensation (probability f(k,θ)) in which the kinetic energy of the atom $\hbar^2k^2/2m$ plus the binding energy L_o (latent heat at the absolute zero, $L_o/k_B = 7.16K$) is completely converted into excitations of the liquid. The excitations produced in (b) and (c) may include quantized surface waves ('ripplons') as well as phonons. Clearly

$$R(k,\Theta) + D(k,\Theta) + f(K,\Theta) = 1 \tag{1}$$

The quantity f(k,θ) is related by the principle of detailed balance to the flux of evaporated atoms with momentum k and angle θ when the liquid is in equilibrium with its vapor at finite temperature. The evaporated flux of atoms with given momentum is exactly balanced by the flux of vapor atoms condensing with the same momentum and direction. The condensing flux is the appropriate Maxwellian ideal-gas expression multiplied by the probability of condensation f(k,θ). The well-known "accommodation coefficient" $\bar{f}(T)$,[17] which is the average probability of condensation, is just f(k,θ) averaged over the Maxwellian distribution at temperature T:

$$\bar{f}(T) = \frac{\iint f(k,\Theta) e^{-\hbar^2 k^2/2mk_B T} k^3 \, dk \sin\Theta \cos\Theta \, d\Theta}{\iint e^{-\hbar^2 k^2/2mk_B T} k^3 \, dk \sin\Theta \cos\Theta \, d\Theta} \tag{2}$$

An important result of our experiment is that $D(k,\theta)$, the probability of inelastic scattering is very small. Within our experimental sensitivity, we could detect no inelastically scattered atoms at all. Making the rather conservative assumption that if inelastic scattering does occur it will be completely diffuse, we were able to establish[10] that $D(k,\theta) \leq 2 \times 10^{-3}$. Since $R(k,\theta)$ is also quite small (see the results in Fig. 3) this means that $f(k,\theta)$ is very close to unity for most k and θ and therefore that the spectrum of evaporated atoms is very close to Maxwellian when the liquid is in equilibrium with the vapor. To give a concrete example, at T = 0.6K integrating over the values of $R(k,\theta)$ in Fig. 3 gives $\bar{f}(0.6) = 0.989 \pm 0.002$ where the uncertainty is mainly due to the uncertainty in $D(k,\theta)$.

The theoretical predictions that we mentioned earlier[12-16] are connected with a possible threshold in the condensation process (c). When $k = 0.5\text{Å}^{-1}$ the combined kinetic energy and latent heat ($\hbar^2k^2/2m + L_0$) of an incident atom is equal to the 'roton' energy 8.65K, that is the energy at the minimum in the well-known phonon spectrum of liquid ^{4}He, shown in Fig. 4. It was predicted[12,13] that at $k = 0.5\ \text{Å}^{-1}$ there would be a discontinuity in the spectrum of evaporated atoms and also in $R(k,\theta)$, since below this momentum there would be insufficient energy to produce a roton excitation. Surprisingly enough there is no sign of this predicted jump in the experimental data and in fact $R(k,\theta)$ seems to depend only on the perpendicular component of the momentum $\hbar k \cos\theta$, rather than on the total momentum $\hbar k$ or kinetic energy $\hbar^2k^2/2m$.

Another interesting feature of the data in Fig. 3 is the fact that $R(k,\theta)$ is so small, less than 5% even for the smallest values of $k \cos\theta$ measured, about $0.05\ \text{Å}^{-1}$. By analogy with the simple quantum mechanical problem of a very slow particle approaching a stepped, attractive potential one might have expected that $R(k,\theta)$ would approach unity as $k \to 0$. At the present time we believe that this will indeed happen, but at values of $k \cos\theta$ much smaller than $0.05\ \text{Å}^{-1}$. This is because of the attractive van der Waals potential outside the liquid. This varies as $-1/z^3$ and writing the Schrodinger equation (for normal incidence) in this region in the form:

$$d^2\Psi/dz^2 + (k^2 + \lambda/z^3)\Psi = 0 \qquad (3)$$

we see that the edge of the potential will not behave like an abrupt step until $k\lambda \ll 1$. For the helium van der Waal's potential, $\lambda = 20\ \text{Å}$ so that $R(k,\theta)$ should approach unity when k is small compared to $\lambda = 0.05\ \text{Å}^{-1}$.

The theoretical interpretation of the experiment has not been completed but we have developed an approximate theory of elastic

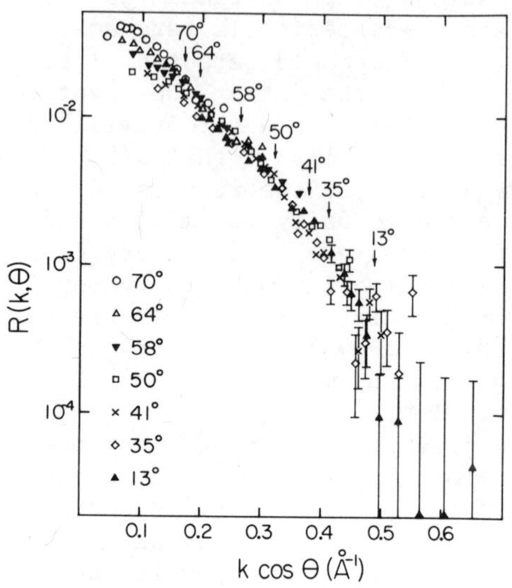

Figure 3. The probability of specular reflection $R(k,\theta)$ for ^4He atoms striking the surface of liquid helium as a function of $\hbar k \cos\theta$, the perpendicular component of their momentum (from Ref. 10). The different symbol shapes refer to the angles of incidence θ shown in the left hand corner. The arrows show the roton threshold ($k = 0.5Å^{-1}$) for each angle. Overlapping points have been omitted for the sake of clarity.

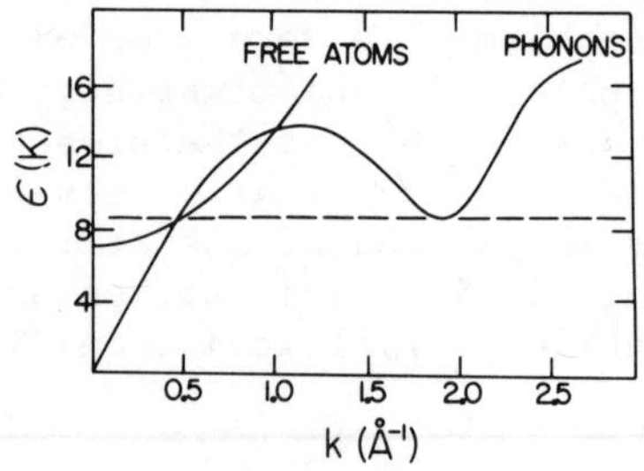

Figure 4. The energy-momentum relation for free atoms $\varepsilon(k)$ relative to their ground-state energy in the liquid, compared with the phonon-roton excitation spectrum. The finite energy for the atoms at k=0 is the binding energy L_0/k_B=7.16K. The energy of an atom is equal to the phonon energy at the roton minimum when $k=0.50Å^{-1}$. (From Ref. 10).

scattering that gives a very satisfactory fit to the data. The approximation is to use the Feynman[18] variational method and to treat the scattered atom as distinguishable from those in the liquid, i.e., the variational wave function is not symmetrized with respect to the scattered atom. The neglect of symmetry is a serious error but perhaps not in the region above the surface where the liquid density is quite low and which we believe has a dominant effect on the reflection coefficient.

The Feynman trial wave function of a state in which one of N ^{4}He atoms has been replaced by an 'impurity' He atom is $\Psi = f(\vec{r}_1)\,\phi(\vec{r}_1 \ldots . \vec{r}_N)$ where ϕ is the ground state for N ^{4}He atoms: $H\phi = 0$ where H is the Hamiltonian and we are now measuring energies from the ground state ($-NL_0$ compared to the vacuum). If the impurity has the same interaction potential but a different mass from ^{4}He, $m' \neq m$, the Hamiltonian becomes $H' = H - \frac{\hbar^2}{2}\left(\frac{1}{m'} - \frac{1}{m}\right)\nabla_1^2$.

When the expectation value of the energy

$$E = \int \Psi^* \, H' \, \Psi \, d\vec{r}_1 \ldots d\vec{r}_N \Big/ \int \Psi^*\Psi \; d\vec{r}_1 \ldots d\vec{r}_N \tag{4}$$

is minimized, Feynman showed that, in bulk helium, $f(\vec{r}_1) = e^{i\vec{K}\cdot\vec{r}_1}$ and the state Ψ has energy $\hbar^2K^2/2m$ above the ground state and momentum $\hbar\vec{K}$. When the theory is applied to helium with a free surface it is convenient to write $f(\vec{r}_1)=\psi(\vec{r}_1)/\sqrt{\rho(\vec{r}_1)}$ where $\rho(\vec{r}_1)$ is the single particle density in the ground state ϕ:

$$\rho(\vec{r}_1) = N \int \phi^2 d\vec{r}_2 \ldots d\vec{r}_N \quad . \tag{5}$$

The probability density for atom 1 in the state Ψ is then $\psi^*(\vec{r}_1)\,\psi(\vec{r}_1)$ (apart from a normalization constant). Minimization of the energy in this case gives

$$\nabla^2 \, \psi(\vec{r}_1) + [E - U(\vec{r}_1)] \;\; \psi(\vec{r}_1) = 0 \tag{6}$$

i.e., a single particle Schrodinger equation with an effective potential $(\hbar^2/2m)U(\vec{r})$ which is related to the particle density $\rho(\vec{r})$ and kinetic energy density $T(\vec{r})$ in the ground state ϕ. This result, which was derived by Lekner,[19] has been applied with considerable success[20-22] to the calculation of the binding energy of ^{3}He to the surface, the potential $U(\vec{r})$ for ^{3}He having a minimum at the surface.

In our scattering problem the 'impurity' is the scattered ^{4}He so that $m'=m$ and the effective potential is given by

$$U(\vec{r}) = a''/a$$

$$a(\vec{r}) = \sqrt{\rho(\vec{r})}$$

$$a''(\vec{r}) = \nabla^2 a(\vec{r}) = \frac{d^2a}{dz^2} \qquad (7)$$

Since $\psi^*\psi$ is the probability density for the scattered atom and ψ obeys a single-particle Schrodinger equation, the current of scattered atoms is conserved and the problem is reduced to finding the single-particle reflection coefficient for the one-dimensional potential $U(\vec{r}) = U(z) = a''/a$. The probability for elastic scattering is thus directly related to the density profile of the liquid $\rho(z) = [a(z)]^2$.

The reader will notice two features of the theory which agree very well with the observations: (a) The predicted reflection coefficient can only depend on the perpendicular component of the incident momentum $\hbar k \cos\theta$. (b) There can be no inelastic scattering, only specular reflection or absorption. It is also clear that the theory can be applied to ^{3}He scattering from ^{4}He although this has not yet been measured.

In calculating the reflection coefficient the proper asymptotic behavior of a(z) outside the liquid must be taken into account. Far from the surface, where the liquid density drops off exponentially, the effective potential must be identical to the real, van der Waals potential:

$$\left.\begin{aligned} a &\to \exp\left[-\left(\beta z + \frac{\lambda}{4\beta z^2}\right)\right] \\ U &\to \beta^2 - \lambda/z^3 \end{aligned}\right\} \quad \text{as } z \to \infty$$

$$\text{where } \hbar^2\beta^2/2m = L_o \ . \qquad (8)$$

When this condition is satisfied it is relatively easy to find analytic forms for a(z) and U(z) which give excellent agreement with experiment. In this way we can use the scattering data to determine the density profile of the liquid at the surface $\rho(z) = [a(z)]^2$. However it is not yet clear how the neglect of symmetry will impair the derived density profile or whether this approximation can be eliminated. The details of the fit and of the theoretical formalism will be reported elsewhere.

This paper was completed while one of us (D. O. Edwards) was on leave at Brookhaven National Laboratory. He would like to acknowledge the hospitality and financial assistance of the Physics

Department there, and particularly considerable assistance from Dr. V. J. Emery in applying the Feynman variational method to the scattering problem.

References

(1) C.-W. Woo, these proceedings.

(2) A. F. Andreev and D. A. Kompaneets, Zh. Eksp. Teor. Fiz. 61, 2459 (1972), (Sov. Phys. JETP 34, 1316 (1972)).

(3) J. R. Eckardt, D. O. Edwards, P. P. Fatouros, F. M. Gasparini and S. Y. Shen, Phys. Rev. Lett. 32, 706 (1974).

(4) D. O. Edwards, S. Y. Shen, J. R. Eckardt, P. P. Fatouros, and F. M. Gasparini, Phys. Rev. B. 1 July, 1975.

(5) J. Bardeen, G. Baym and D. Pines, Phys. Rev. 156, 207 (1967).

(6) C. Ebner and D. O. Edwards, Physics Reports 2 C, 77 (1971).

(7) Y. Disatnik and H. Brucker, J. Low Temp. Phys. 7, 491 (1972).

(8) H. M. Guo, D. O. Edwards, R. E. Sarwinski and J. T. Tough, Phys. Rev. Lett. 27, 1259 (1971).

(9) See for instance, W. C. Black, A. C. Mota, J. C. Wheatley, J. H. Bishop and P. M. Brewster, J. Low Temp. Phys. 4, 391 (1971).

(10) D. O. Edwards, P. Fatouros, G. G. Ihas, P. Mrozinski, S. Y. Shen, F. M. Gasparini, C. P. Tam, Phys. Rev. Lett. 34, 1153 (1975).

(11) J. R. Eckardt, D. O. Edwards, F. M. Gasparini and S. Y. Shen, Proc. of 13th Intern. Conf. on Low Temp. Physics, LT-13, ed. Timmerhaus, O'Sullivan and Hammel, (Plenum Press, New York, 1974), p. 518.

(12) A. Widom, Phys. Letters 29 A, 96 (1969); D. S. Hyman, M. O. Scully and A. Widom, Phys. Rev. 186, 231 (1969).

(13) P. W. Anderson, Phys. Letters 29 A, 563 (1968).

(14) A. Griffin, Phys. Letters 31 A, 222 (1970).

(15) J. I. Kaplan and M. L. Glasser, Phys. Rev. A 3, 1199 (1971).

(16) M. W. Cole, Phys. Rev. Lett. 28, 1622 (1972).

(17) See for instance, K. R. Atkins, R. Rosenbaum and H. Seki, Phys. Rev. 113, 751 (1959); G. H. Hunter and D. V. Osborne, J. Phys. C: Proc. Phys. Soc., London 2, 2414 (1969); D. G. Blair, Phys. Rev. A 10, 726 (1974).

(18) R. P. Feynman, Phys. Rev. 94, 262 (1954).

(19) J. Lekner, Phil. Mag. 22, 669 (1970).

(20) W. F. Saam, Phys. Rev. A 4, 1278 (1971).

(21) Y. M. Shih and C.-W. Woo, Phys. Rev. Lett. 30, 478 (1973).

(22) C. C. Chang and M. Cohen, Phys. Rev. A 8, 3131 (1973).

THERMAL EXPANSION COEFFICIENT AND UNIVERSALITY NEAR THE SUPERFLUID TRANSITION OF ^{4}He UNDER PRESSURE

K. H. Mueller and F. Pobell

Institut für Festkörperforschung Kernforschungsanlage

517 Jülich, W. Germany

and

Guenter Ahlers

Bell Laboratories

Murray Hill, New Jersey 07974

Summary: Recent measurements[1] of the specific heat C_p of pressurized liquid helium near the λ-line gave results which seemed to be in conflict with the concept of universality;[2] specifically the amplitude ratio A/A' of the singular part of C_p above and below T_λ appeared to depend upon the pressure P.[1] We report on new determinations of A/A' for ^{4}He near T_λ based upon measurements of the isobaric thermal expansion coefficient β_p,[3] which is a linear function of C_p sufficiently near T_λ. Our measurements were made by a new technique, to be described in detail elsewhere,[3] which is characterized by high pressure stability ($\Delta P/P \lesssim 2\cdot10^{-8}$), and temperature resolution ($\Delta T/T \approx 2\cdot10^{-7}$). Our experiment covered the ranges 5 bar $\leq P \leq$ 29 bar, and $2\cdot10^{-5} \leq |t| \leq 0.07$, where $t = T/T_\lambda - 1$. The data were fit to

$$\beta_p = (A/\alpha)\ t^{-\alpha}\ [1 + D\ t^x] + B$$

for $T > T_\lambda$, and simultaneously to the same function with primed coefficients for $T < T_\lambda$.

Using data for $|t| \leq 0.003$, and imposing the constraints $\alpha=\alpha'$, $x = x' = 0.5$, $B = B'$ we obtain the pressure independent (universal) values

$$\alpha = \alpha' = -0.026 \pm 0.004$$

$$A/A' = 1.11 \pm 0.02$$

where the uncertainties are standard errors. The exponent is more accurate than (but in excellent agreement with) the result $\alpha = \alpha' = -0.02 \pm 0.02$ derived from C_p.[1] Contrary to the conclusions based on the specific heat,[1] we conclude that A/A' is universal along the λ-line in pressurized ^{4}He, consistent with theoretical predictions.[2] Details can be found in Ref. 3.

References

(1) G. Ahlers, Phys. Rev. A8, 530 (1973).

(2) See, for instance, L. P. Kadanoff, in Critical Phenomena, Proceedings of the International School Enrico Fermi, Course Ll, edited by M. S. Green (Academic Press, N.Y., 1971); R. B. Griffiths, Phys. Rev. Lett. 24, 1479 (1970); D. Jasnow and M. Wortis, Phys. Rev. 176, 739 (1968); E. Brezin, J. C. Guillou, and J. Zinn-Justin, Phys. Lett. 47A, 285 (1974).

(3) K. H. Mueller, F. Pobell, and G. Ahlers, Phys. Rev. Lett. 34, 513 (1975), and to be published in Phys. Rev.

BOGOLIUBOV'S COMPENSATION OF DANGEROUS DIAGRAMS AND THE EXCITATION SPECTRUM OF He II*

Donald H. Kobe

Department of Physics, North Texas State University

Denton, Texas 76203

Abstract: A microscopic theory of boson systems was first given by Bogoliubov in 1947. A decade later he extended the theory and proposed the *principle of compensation of dangerous diagrams* (PCDD) to determine the coefficients in a canonical transformation to quasiparticles. In the first part of this paper I review the present status of the PCDD. The second part of the paper is a calculation of the excitation spectrum of He II done with G. W. Goble based on the compensation of the lowest order dangerous diagrams which gives the Hartree-Fock-Bogoliubov (HFB) theory. When two phenomenological modifications to the HFB theory are made it can be applied to He II with a realistic potential, and qualitative agreement with the spectrum at low temperature is obtained.

I. COMPENSATION OF DANGEROUS DIAGRAMS

In his remarkable 1947 paper[1] Bogoliubov made two approximations on the Hamiltonian to obtain the energy spectrum for a weakly interacting boson system. First, he replaced the creation and annihilation operators $a_o^\dagger$ and a_o, respectively, which satisfy the boson commutation relations, by the c-number $N_o^{1/2}$, where N_o is the number of particles in the zero momentum state. The neglect of terms with three or four operators gives a quadratic Hamiltonian which can be diagonalized by making a canonical transformation to quasiparticles (QP). The coefficients in the canonical transformation

---

*Work supported in part by the North Texas State University Faculty Research Fund.

were chosen to eliminate all but the diagonal term in the Hamiltonian. The result of these approximations is the familiar energy spectrum

$$E_k = [k^2/2m \quad (k^2/2m + 2\rho_0 V_k)]^{1/2} \tag{1}$$

where E_k is the energy of a QP with momentum k, m is the mass of the boson, V_k is the Fourier transform of the two-body potential, and ρ_0 is the density of particles in the zero momentum state.

With the advent of a microscopic theory of superconductivity[2] in 1957, Bogoliubov proposed the principle of compensation of dangerous diagrams (PCDD) for both boson[3] and fermion systems.[3,4] Since the process in which a pair of QP is created from the vacuum is unphysical and results in divergences in the ground-state energy, Bogoliubov proposed that the sum of all diagrams which lead from the vacuum to a two QP state be set equal to zero. However, setting the first-order diagram equal to zero can be justified by the energy variational principle.[5] The question then arose as to whether the full PCDD should be used or just the first-order term in the PCDD. The question became a practical one when Woo and Ma[6] showed in the theory of a charged Bose gas that higher-order dangerous diagrams were indeed required to eliminate divergences.

The justification of the PCDD was then made on the basis of two different variational principles: (1) the overlap between the QP vacuum state and the true ground state is a maximum,[7] and (2) the expected number of QP in the true ground state is a minimum.[8] This latter variational principle has been used in finite fermion systems[9] to obtain Löwdin's natural spin orbitals[10] which diagonalize the single-particle density matrix. For boson systems, a new class of "dangerous diagrams" is obtained when the zero-momentum state is treated exactly. The sum of all the diagrams that lead from the vacuum to the single QP state is also shown by the variational principle to be equal to zero.[7,8]

In extending the theory of boson systems to finite temperatures, the question naturally arose as to whether the PCDD could also be extended. We have shown that minimizing the expected number of QP calculated in the grand canonical ensemble results in a finite temperature version of the PCDD.[11] If only the lowest-order diagram is used the result is the same as minimizing the Helmholtz free energy calculated from a single QP Hamiltonian, which gives the Hartree-Fock-Bogoliubov (HFB) theory at finite temperature. The higher-order dangerous diagrams at finite temperature have not yet been investigated.

II. EXCITATION SPECTRUM OF HE II

We have recently used the Hartree-Fock-Bogoliubov (HFB) theory at finite temperature (compensation of lowest order dangerous diagrams) to calculate the energy spectrum of He II.[12] Even though the excitation spectrum has been studied by many others, we felt a reinvestigation was warranted because (1) new ^{4}He interatomic potentials, fit to various data on the gaseous phase, have become available,[13] (2) the temperature dependence of the spectrum has not been studied for a realistic potential, and (3) a local t-matrix was shown to emerge in a very natural way from the HFB theory when pairing correlations are completely taken into account.[14]

However, the HFB theory immediately runs into trouble. The first problem is that, contrary to experiment and the Hugenholtz-Pines theorem,[15] an energy gap[16] appears in the spectrum. For strongly repulsive potentials this gap could be extremely large. The chemical potential can easily be chosen to eliminate the gap, but then the variational principle applied to the density amplitude for the zero-momentum state must be abandoned. In other words, the lowest-order diagram leading from the vacuum to the single QP state is not compensated.

Another difficulty is now encountered. The single particle energy $k^2/2m$, has Hartree-Fock-Bogoliubov single-particle terms added to it. For a potential which is very strongly repulsive at short distances, these terms can diverge. To avoid this divergence, the single particle energy can be phenomenologically modified so it becomes $k^2/2m^*$, where m^* is an effective mass.

The energy spectrum then becomes[12]

$$E_k = [k^2/2m^* \quad (k^2/2m^* + 2\Delta_k)]^{1/2} \tag{2}$$

When the lowest-order dangerous diagram which leads from the vacuum to the two QP state is set equal to zero, the term Δ_k is given by the integral equation

$$\Delta_k = \rho_0 V_k - (2\pi)^{-3} \int d^3p \, V_{k-p} \, (\Delta_p/2E_p) \coth(1/2 \, \beta E_p) \ , \tag{3}$$

where ρ_0 is the density of particles in the zero-momentum state. Equation (3) results from considering a pair of particles of equal and opposite momentum excited from the zero-momentum condensate to the states $\vec{k}$ and $-\vec{k}$, which can then scatter with each other an arbitrary number of times. It is convenient to eliminate ρ_0 from Eq. (3) by defining a t_k such that

$$\Delta_k = \rho_0 \, t_k \tag{4}$$

When divided by $\rho_o \neq 0$, Eq. (3) gives an integral equation,

$$t_k = V_k - (2\pi)^3 \int d^3p \, V_{k-p} \, (t_p/2E_p) \coth(1/2 \, \beta E_p) \, , \qquad (5)$$

for a local, temperature-dependent t-matrix which has a completely dressed energy denominator $2E_p$. The first calculation based on a t-matrix[17] inserted it ad hoc into Eq. (1). The t-matrix in Eq. (5) arises naturally here because pairing correlations are completely taken into account.

Since the two-body potential V(r) is strongly repulsive at short distances, its Fourier transform may not exist or may be difficult to calculate. Thus it is very convenient to Fourier transform Eq. (5) to give

$$t(r) = V(r) - V(r) \int_0^\infty dr' \, K(r,r') \, t(r') \, , \qquad (6)$$

where the kernel in the integral is defined as

$$K(r,r') = (r'/\pi r) \int_0^\infty dp \;\; E_p^{-1} \sin pr \sin pr' \coth(1/2 \, \beta E_p). \qquad (7)$$

The local t-matrix in configuration space t(r), also called the reaction operator, is the Fourier transform of t_k, so

$$t_k = 4\pi \int_0^\infty dr \, r^2 \, [\sin kr/kr] \, t(r) \, . \qquad (8)$$

The density of particles in the zero-momentum state ρ_o is the total density ρ of the system minus the density of particles not in the zero momentum state which gives

$$\rho_o = \rho - (2\pi)^{-2} \int_0^\infty dp \, p^2 \, [(p^2/2m^* + \rho_o t_p) E_p^{-1} \coth(1/2 \, \beta E_p) - 1] \, , \qquad (9)$$

where E_p is determined from Eq. (2) with Eq. (4) substituted into it.

The method of calculation is first to assume an energy E_k. Then the kernel in Eq. (7) is calculated. Equation (6) for the reaction operator t(r) is solved, and then Eq. (8) is used to obtain t_k. The t_k is used in Eq. (9) to calculate ρ_o which is then substituted into Eq. (4) and Eq. (2). The process is then repeated until self-consistency is reached. For simplicity we have used $m^* = m$.

For the Morse-dipole-dipole-2 (MDD2) potential,[13] the result of the calculation of the reaction operator t(r) is shown in Fig. 1 for several temperatures. For large interparticle distance r the reaction operator approaches the potential. However, for small

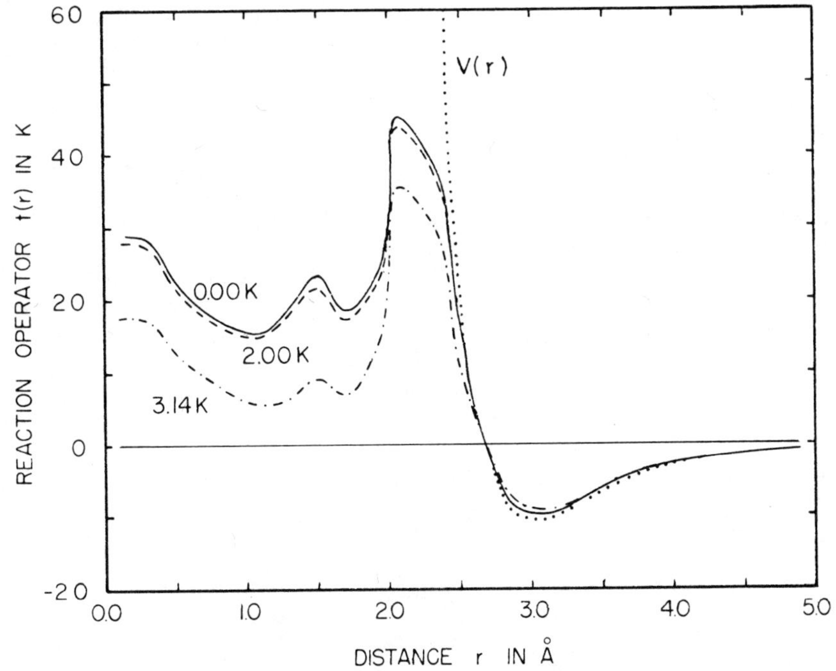

Figure 1. The reaction operator t(r) as a function of the interatomic distance calculated from the Morse-dipole-dipole-2 (MDD2) potential (dotted curve) of Ref. 13 at temperatures of 0.00 K (solid curve), 2.00 K (dashed curve), and 3.14 K (dot-dashed curve).

inter-particle distances the reaction operator t(r) is much less repulsive, because it includes some of the two-body correlation which is present in the true wave function. The excitation spectrum corresponding to the t(r) in Fig. 1 is shown in Fig. 2 for the same temperatures. For the experimental density ρ of 0.0219 atoms $\mathring{A}^{-3}$, the condensate fraction ρ_0/ρ varies smoothly from about 60% at 0.00 K to about 40% at 3.14 K. Experimentally, the condensate fraction has been determined to be 2.4 ± 1%, which is much smaller.[18]

There is only qualitative agreement between the experimental points of Cowley and Woods[19] at 1.12 K in Fig. 2 and the theoretical spectrum. As the temperature increases, the calculated spectrum approaches the free particle spectrum, which is contrary to experiment. The poor agreement between experiment and theory can be ascribed to the neglect of QP interactions (or terms involving three and four particles). To include these terms would involve using the PCDD in higher order in conjunction with the QP self-energy. The self-consistent solution of the resulting equations would indeed be

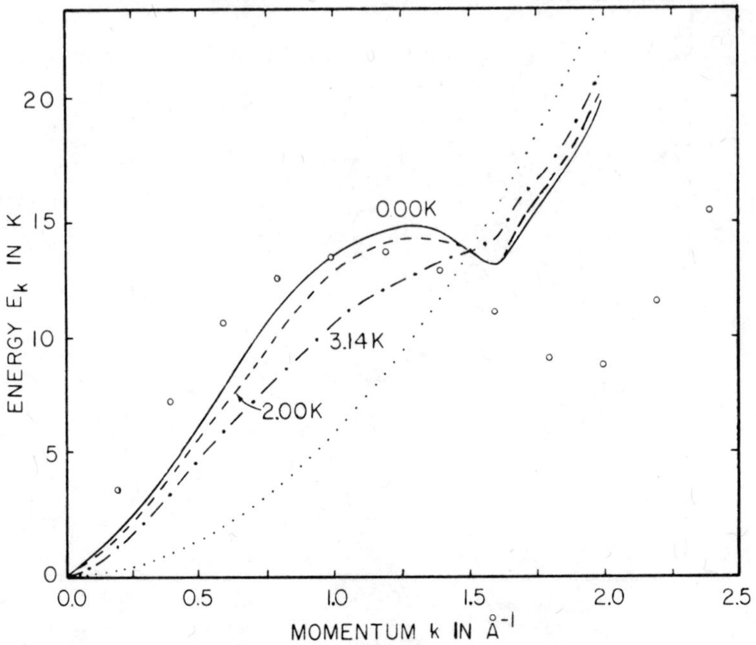

Figure 2. The excitation spectra E_k as a function of momentum k calculated from the MDD2 potential for the temperatures 0.00 K (solid curve), 2.00 K (dashed curve), and 3.14 K (dot-dashed curve). The free particle spectrum is given by the dotted curve. The open circles are the experimental data of Cowley and Woods (Ref. 19) at 1.12 K.

a formidable calculation.

However, there are still a number of modifications which can be made in the HFB theory. The first is to vary the effective mass until better agreement is attained. Another modification is to replace ρ_0 in Eqs. (3) and (4) by the total density , which would simplify the set of equations by removing Eq. (9). This modification arises naturally in the theory of Sunakawa, et al.[20] of the excitation spectrum based on the density and velocity operators. In the HFB theory this replacement should be justified on the basis of perturbation theory.

In order for the theory used here to be completely satisfactory, it is necessary to justify from fundamentals (1) the choice of the chemical potential to eliminate the energy gap, and (2) the use of the effective mass approximation. Because of the mathematical ambiguities in the Sunakawa theory,[21] it seems worthwhile to extend the theory presented here as far as possible.

References

(1) N. N. Bogoliubov, J. Phys. USSR 11, 23 (1947).

(2) J. Bardeen, L. N. Cooper, and J. R. Schrieffer, Phys. Rev. 108, 1175 (1957).

(3) N. N. Bogoliubov, V. V. Tolmachev, and D. V. Shirkov, A New Method in the Theory of Superconductivity (Academy of Sciences of the USSR Press, Moscow, 1958), Chap. 1 (English translation: Consultants Bureau, New York, 1959): Fortshr. Physik 6, 605 (1958).

(4) N. N. Bogoliubov, Zh. Eksp. Teor. Fiz. 34, 58 (1958) [Sov. Phys. -JETP 7, 41 (1958)]; Nuovo Cimento 7, 794 (1958).

(5) A. Coniglio and M. Marinaro, Nuovo Cimento 48, 249 (1967).

(6) C. W. Woo and S. K. Ma, Phys. Rev. 159, 176 (1967).

(7) D. H. Kobe, J. Math. Phys. 9, 1779 (1968).

(8) D. H. Kobe, J. Math. Phys. 9, 1795 (1968).

(9) D. H. Kobe, J. Chem. Phys. 50, 5183 (1969).

(10) P.-O. Löwdin, Phys. Rev. 97, 1474 (1955).

(11) D. H. Kobe and G. W. Goble, J. Math. Phys. 15, 1835 (1974).

(12) G. W. Goble and D. H. Kobe, Phys. Rev. A10, 851 (1974).

(13) L. M. Bruch and I. J. McGee, J. Chem. Phys. 52, 5884 (1970).

(14) D. H. Kobe, Ann. Phys. (N.Y.) 47, 15 (1968).

(15) N. M. Hugenholtz and D. Pines, Phys. Rev. 116, 489 (1959).

(16) M. Girardeau and R. Arnowitt, Phys. Rev. 113, 755 (1959).

(17) K. A. Brueckner and K. Sawada, Phys. Rev. 106, 1117, 1128 (1957).

(18) H. A. Mook. R. Scherm, and M. K. Wilkinson, Phys. Rev. A6, 2268 (1972).

(19) R. A. Cowley and A. D. B. Woods, Can. J. Phys. 49, 177 (1971).

(20) S. Sunakawa, S. Yamasaki, and T. Kebukawa, Prog. Theor. Phys. (Kyoto) 41, 919 (1969).

(21) D. H. Kobe and G. C. Coomer, Phys. Rev. A7, 1312 (1973).

ELECTRON-HOLE DROPLETS IN GERMANIUM AND SILICON

Monique Combescot

Laboratory of Atomic and Solid State Physics,
Cornell University
Ithaca, New York 14853

Abstract: The electron hole plasma is a stable steady-state in germanium and silicon. Its thermodynamical properties such as energy, density and phase diagram are studied experimentally and theoretically.

The electron gas is the most simple quantum fluid one can imagine. Various properties which must exist in a quantum fluid have been studied on it. But, unfortunately, the electron gas exists in a metal where the background of the positively charged ions has to be taken into account. As ions are more complex than electrons, theories in metal, which have to include ions, are not in a favorable situation because of the lack of knowledge of them.

When a new state, supposed to be an electron hole plasma, had been discovered in germanium, theorists were enthusiastic because they could test their theories in an (quasi) ideal situation. Germanium has been greatly studied and its band structure (even if it is not very pleasant to include it) is well known. Germanium also can be made very clean so that a lack of agreement between theory and experiment will have to be directly related to the quantum fluid theory without any possible escape in ions potential, lattice, etc.

By shining light on a semiconductor, nonequilibrium carriers are created. Usually, the electron and hole (e-h) bind together in order to form excitons. Looking at the luminescence due to the recombination of electrons (e) and holes (h), one gets the usual exciton lineshape $I(\omega)\ \alpha \sqrt{\omega}\ \exp\ (-\omega/kT)$. In germanium and silicon, the luminescence spectrum presents another line at lower energy[1] than the excitonic one (Fig. 1) The lineshape of this new line fits surprisingly well with the recombination of free electrons and free holes in the conduction and valence band

$$I_{EHD}(\omega) \; \alpha \int_{0}^{\mu_e} d\omega_e \sqrt{\omega_e} \int_{0}^{\mu_h} d\omega_h \sqrt{\omega_h} \; \delta(\omega - \omega_e - \omega_h).$$

Keldysh suggested that this new state might be an e-h plasma (the two components having here light masses). If one looks at the energy $\bar{E}$ of a pair e-h as a function of r_s (interparticle distance expressed in Bohr radius) at zero temperature, for small density, or large r_s, $\bar{E}$ goes to the exciton binding energy E_{FE}. At very high density, $\bar{E}$ is controlled by the kinetic energy and diverges; at $r_s \sim 1$, $\bar{E}$ has, in general, a minimum E_o. Let n_o be the corresponding density. If this minimum is lower than the exciton energy, the ground state of a system of e-h is a plasma (Fig. 2a). As the initial average density n of e-h is usually lower than n_o, one gets at T = 0 drops of plasma in equilibrium with vacuum. If $E_o > - E_{FE}$, the ground state is an exciton gas (Fig. 2b).

In order to check Keldysh's suggestion theoretically, the calculation of the energy of a two-component plasma at T = 0 has been made by various authors: Combescot-Nozieres[2] and Brinkman-Rice[3] have obtained the correlation energy using the RPA approximation "improved" by Nozieres-Pines, or Hubbard. Vashista-Bhattarcharyya-Singwi[4] have applied the Singwi-Tosi-Land Sjolander method which uses a set of self-consistent equations for the particle-particle correlation function and takes into account some

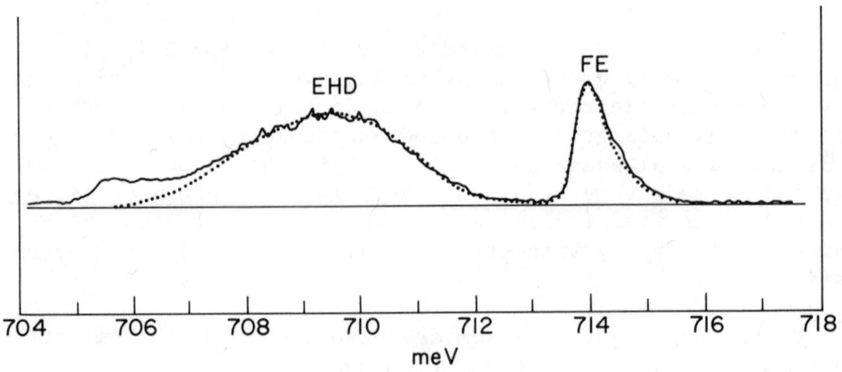

Figure 1. Luminescence spectrum of Ge (from Reference 11) solid line: experiment; dotted line; theory (including slit broadening).

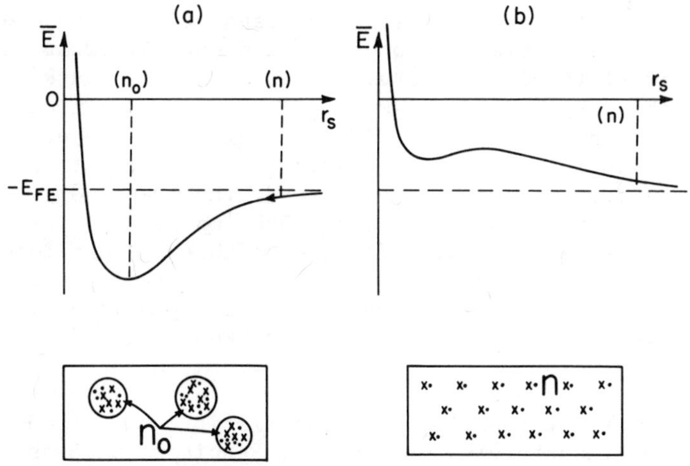

Figure 2. Energy of an e-h pair as a function of r_s for T = 0 a) when the plasma is stable, b) when the exciton gas is stable.

type of diffusion processes neglected in RPA.

In the case of Ge, where the binding energy of excition is of order 3.6 mev, the Hartree-Fock approximation for the energy gives 1.8 mev; but adding the correlation contribution, one gets 5.6 mev (within 10% for the various theories) so that the binding energy of the plasma with respect to the exciton energy is of order 2 mev. (Experimental values[5] are quoted between 1.4 mev and 2 mev. Note that the uncertainty on the binding energy of the exciton affects directly the theoretical value of ϕ.) The equilibrium density n_o of the plasma is 2×10^{17} part/cm^3. For silicon, $\phi \simeq 6$ mev and $n_o \simeq 3 \times 10^{18}$ part/cm^3.

The band structure is very important to determine if the plasma is stable or not. With only one valence band and one conduction band with same mass, the energy $\bar{E}$ is 0.9 E_{FE}. The most important characteristic[6] of Ge, which makes the plasma stable, is its multivalley band structure (that decreases the kinetic energy, and increases correlations).

For $T \neq 0$, excited states different from the e-h plasma can be thermically populated, and so the plasma is in equilibrium with a gas a priori composed of excitons, biexcitons, and free carriers. For small T, the gas has a small density and can be approximated by a perfect gas. Writing equilibrium between all the species, one can show[7] that the gas is mostly composed of excitons. When T increases, free carriers become more and more important and, also, excitons start to disassociate. Thomas, Hensel, and Rice[8] have measured the e-h phase diagram (Figure 3). The critical point is found at $T_c \simeq 6.7°K$ and $n_c \simeq 0.8\ 10^{17}$ part/cm^2, and agrees reasonably well with the theoretical estimate by Combescot[6] and Vashista[9] based on an expansion in temperature of the free energy ($T_c \simeq$ 6-8°K and $n_c \simeq 0.8\ 10^{17}$ part/cm^3). One important point to note is is that the critical density is high so that the excitons are mostly dissociated. The shape of the phase diagram using a phenomenological model based on a spin 1 lattice gas model in order to represent the two-component plasma has been obtained by Droz and Combescot.[10]

One important question remains: is it a Mott transition in the gas? At low temperature, e-h bound in excitons make an insulating gas; near T_c, excitons are dissociated and so the gas is a conductor. A real Mott transition is not expected because it will be smoothed by temperature broadening, but some anomaly in the phase diagram may remain. The problem is that the region where the Mott transition may occur is very hard to investigate experimentally and theoretically.

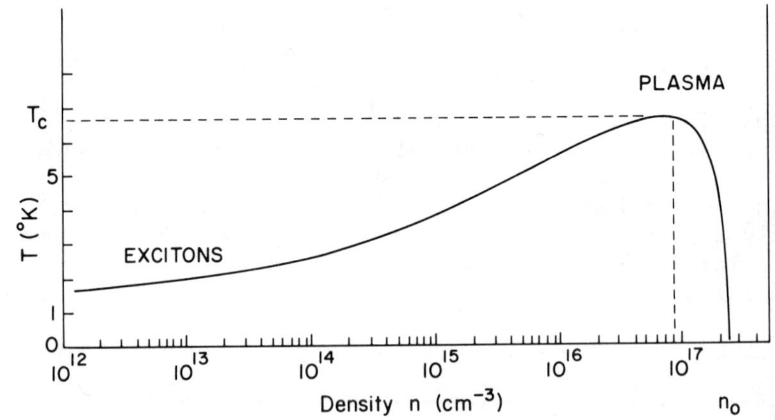

Figure 3. e-h phase diagram for Ge.

References

(1) C. Benoit a la Guillaume, M. Voos, and F. Salvan, Phys. Rev. B5, 3079 (1972); Phys. Rev. B7, 1723 (1973); G. A. Thomas, T. G. Phillips, T. M. Rice, and J. C. Hensel, Phys. Rev. Lett. 31, 386 (1973).

(2) M. Combescot and P. Nozieres, J. Phys. C5, 2369 (1972).

(3) W. F. Brinkman and T. M. Rice, Phys. Rev. B7, 1508 (1973).

(4) P. Vashista, P. Bhattacharyya, and K. S. Singwi, Phys. Rev. Lett. 30, 1248 (1973).

(5) Y. E. Pokrovskii, Phys. Stat. Solidi (a) 11, 385 (1972); C. Benoit a la Guillaume, and M. Voos, Solid State Commun. 12, 1257 (1973); J. C. Hensel, T. G. Phillips, and T. M. Rice, Phys. Rev. Lett. 30, 227 (1973).

(6) M. Combescot, Phys. Rev. B10, 5045 (1974).

(7) M. Combescot, Phys. Rev. Lett. 32, 15 (1974).

(8) G. A. Thomas, T. M. Rice, and J. C. Hensel, Phys. Rev. Lett. 33, 219 (1974).

(9) P. Vashista, G. D. Shashikala, and K. S. Singwi, Phys. Rev. Lett. 33, 911 (1974).

(10) M. Droz and M. Combescot, preprint.

(11) T. K. Lo, B. J. Feldman, and C. D. Jeffries, Phys. Rev. Lett. 31, 224 (1973).

SUPERFLUIDITY IN NEUTRON STARS*

Gordon Baym

Department of Physics, University of Illinois

Urbana, Illinois 61801

Abstract: Possible states of superfluidity in the liquid interiors of neutron stars, including neutron pairing, proton pairing and pion condensation are described, and recent work on the subject is reviewed.

I. INTRODUCTION

Neutron stars are very tiny objects with masses on the order of that of the sun and radii on the order of 10 kilometers. Typical densities in the interior are $\sim 10^{14-15}$ g/cm^3, comparable to the density $\sim 3\times10^{14}$ g/cm^3 inside large nuclei. The outer kilometer or so of a neutron star is a crust of normal metal permeated at higher densities by a degenerate quantum liquid of neutrons, as well as a free relativistic electron liquid. The crust surrounds a liquid interior composed primarily of a neutron quantum liquid mixed with a smaller fraction of degenerate proton and electron liquids in electrical neutrality. Matter at such very high densities has the possibility of existing in some rather unusual states.[1] In this paper I would like to review possible states of superfluidity in neutron stars. These include states of neutron pairing and of proton pairing, in which one expects certain interesting variations from laboratory superfluids and superconductors, and a possible state of "pion condensation," involving the spontaneous formation, in the matter, of charged pions; because pions are bosons the matter

---

*Research supported in part by U. S. National Science Foundation Grant NSF GP-40395.

in such a state is a charged Bose superconductor, in which one has a pairing of neutron-particles with proton-holes. Let us first consider the BCS-like pairing states.

II. NEUTRON AND PROTON PAIRING

The neutrons in the crust and interior of neutron stars, as well as the protons in the interior are most likely superfluid. It is very unlikely that the electrons will ever by superconducting. One can estimate, from the BCS weak coupling theory, the transition temperature $T_{c,e}$ for electron superconductivity to be $T_{c,e} \sim T_{f,e} e^{-1/N(0)V}$; $T_{f,e}$ is the electron Fermi temperature expressed in terms of the electron Fermi momentum p_e by $kT_{f,e} = cp_e$ for fully relativistic electrons. The density of states (for one spin) at the electron Fermi surface is $N(0) = p_e/2\pi^2\hbar^3c$, since the electron effective mass is p_e/c. The mean net attraction V between electrons is a quantity of order $e^2(\hbar/p_e)^2$, so that $N(0)V \sim e^2/\hbar c$. Thus, $T_{c,e}$ is essentially zero; at any realistic T the electrons are normal.

Superfluidity of neutrons, as well as protons, arises from pairing interactions, as in laboratory superconductors. Estimates of neutron and proton energy gaps for $^1S_0$ pairing have been derived microscopically by many authors using a variety of nucleon-nucleon potentials.[2] These authors' results for the neutron $^1S_0$ gaps are in reasonable agreement with the early calculations of Hoffberg et al.[3] using effective pairing interactions derived from phase shifts. At the relative neutron-neutron scattering energies in neutron star matter above nuclear densities, the $^3P_2$ is the dominant attractive phase shift due to the strong tensor force, and thus $^3P_2$ pairing should be favored at such densities.[3] Detailed calculations of anisotropic $^3P_2$ pairing in neutron matter are given in Ref. 4. It should be pointed out though that all these calculations should be regarded as first estimates. Because the gaps depend exponentially on the strength of the interaction, the results depend sensitively on small effects such as modification of the proton-proton interaction due to polarization of the neutron liquid, and single particle self-energies and lifetimes.

Figure 1 shows the neutron transition temperatures computed by Hoffberg et al., together with the $^1S_0$ proton transition temperature computed by Chao et al. The superfluid transition temperatures for neutron and for proton pairing may be estimated from the BCS result $T_c \simeq \Delta/1.76$ for $^1S_0$ pairing, where Δ is the corresponding T=0 energy gap. For the $^3P_2$ gap computed in Ref. 3, $T_c \simeq \Delta/2.4$, where Δ is the maximum gap, as a function of angle on the Fermi surface. The transition temperatures are generally well above the

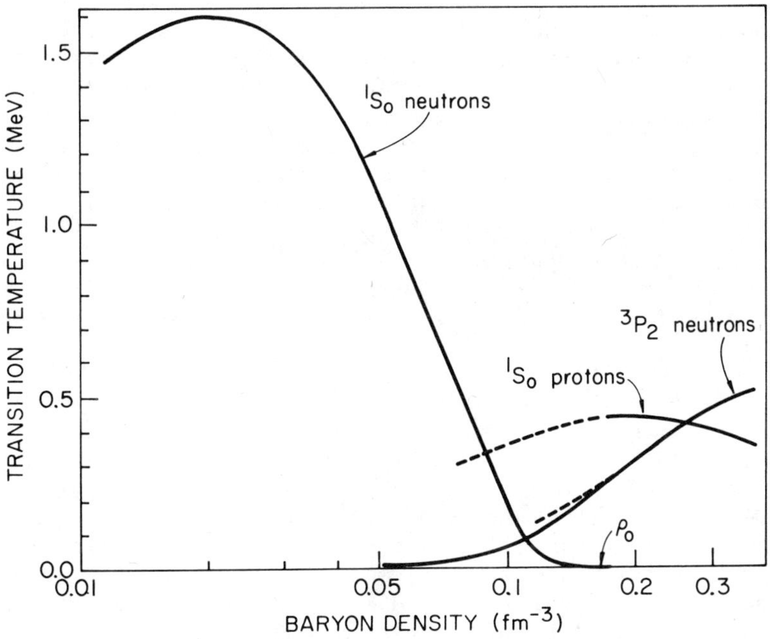

Figure 1

expected ambient temperatures in all but very young neutron stars. The picture of superfluidity in neutron stars that emerges is that the neutron fluid in the crust should be paired in $^1S_0$ states. The transition from $^1S_0$ to $^3P_2$ pairing occurs (coincidentally) at about the density where the nuclei dissolve, and thus in the liquid interior one expects $^3P_2$ neutron pairing, and $^1S_0$ proton pairing.

One consequence of neutron superfluidity is that because neutron stars rotate, the neutrons should form an array of quantized vortices, analogous to the vortices in rotating superfluid helium II. The quantum of circulation is h/2m, where 2m is the mass of a pair of neutrons. For a rotation period of 1/30 sec (that of the Crab pulsar) the vortex spacing is $\sim 10^{-3}$ cm, small compared with characteristic stellar dimensions. By contrast, the vortices are widely spaced compared with the neutron coherence length $\xi_n = \hbar^2 k_n/\pi\Delta_n m \sim 10^{-12}$ cm. Here Δ_n is the neutron energy gap, and k_n the neutron Fermi wavenumber. The number of vortices present is always large enough that the moment of inertia of the neutron superfluid, in the case of $^1S_0$ pairing, is not reduced from the classical value. In general, a $^3P_2$ superfluid is anisotropic with complicated flow properties. Ginzburg-Landau equations for describing these properties are given in Refs. 5 and 6.

The vortex lattice in a neutron superfluid has a mode of oscillation, discussed by Tkachenko,[7] in which the lines are displaced parallel to themselves. The propagation velocity of this mode, for a triangular lattice, is $c_r = (\hbar\Omega/8m)^{1/2} \sim 0.1$ cm/sec, where Ω is the rotation frequency. This mode is essentially the limit, in a quantum fluid, of the "2Ω" inertial mode of a rotating classical fluid,[8] as the wave vector becomes orthogonal to the rotation axis. The coupling of the neutron superfluid to the changed particles, and hence to the outside world, takes place via excitation of Tkachenko modes. Detection and study of their properties in liquid He II would be quite useful to understanding the dynamics of pulsars.

Let us consider now the properties of proton superconductivity. The critical magnetic field H_c (or H_{c1} for type II superconductivity) is of order 10^{16} gauss.[9] (Note that the scale of critical fields is $\sim(hc/2e)/4\pi r_p^2$, where r_p is the mean proton separation; for $r_p \sim 10^{-13}$ cm, this is $\sim 10^{17}$ gauss.) Since $B \sim 10^{12}$ gauss in a neutron star one would normally expect a Meissner effect in which the magnetic flux is expelled from the regions of superconducting protons. Such an expulsion would have a serious effect on the electromagnetic properties of pulsars. However, because of the enormous electrical conductivity of the normal state[10] ($\sigma \sim 10^{29}$ sec^{-1}) the time[11] $\tau \simeq (4\pi\sigma/c^2)R^2(B/2H_c)$ required to expel flux from a region of size $R \sim 1$ km is $\gg 10^8$ years. The superconductivity, rather than waiting so long, simply nucleates with the field present. While this is technically a metastable situation, the actual lifetime of the state can be shown to be orders of magnitude longer than the time τ.

If the protons are a type I superconductor they will learn to live with the field by forming an intermediate state configuration, with alternate layers of superconducting material which is field free, and normal material containing the field. On the other hand, for type II superconductivity, which is more likely, the magnetic flux will be contained in an array of quantized vortices, each of a single flux quantum $\phi_0 = hc/2e$. This is the same state as occurs in laboratory type II superconductors for $H_{c1} < B < H_{c2}$, only now $B \ll H_{c1}$.

The criterion for type II superconductivity is that the penetration depth λ should be greater than ξ_p, the proton coherence length, divided by $\sqrt{2}$. Estimating λ from the London result $\lambda = (m c^2/4\pi n_p e^2)^{1/2}$, where n_p is the proton number density, and using $\xi_p = \hbar^2 k_p/\pi\Delta_p m$, where k_p is the proton Fermi wavenumber, we find

$$\frac{\xi_p}{\sqrt{2}\,\lambda} = \left(\frac{8}{3\pi^2}\frac{e^2}{\hbar c}\frac{\hbar k_p}{m\,c}\right)^{1/2} \frac{\varepsilon_p}{\Delta_p} \; ; \qquad (1)$$

here $\varepsilon_p = h^2k_p^2/2m$. From the calculations of Chao et al.,[2] ε_p/Δ_p is ~ 10 from the outer surface of the liquid interior to $\rho \sim 5\times 10^{14}$ g/cm^3. In this range $\xi_p > \sqrt{2}\lambda$, indicating that the protons should be a type II superconductor.

III. PION CONDENSATION

The question of whether pions appear in dense neutron star matter is of substantial theoretical interest, because of their possible effects on the neutrino cooling of neutron stars and the equation of state, as well as the state of superfluidity to which they lead. Neglecting for the moment the strong interactions of pions with the matter, one sees that π^- will form via $n\rightarrow p+\pi^-$ once the neutron-proton chemical potential difference $\mu_n-\mu_p$ exceeds the π^- rest mass $m_\pi c^2 = 139.6$ MeV. In neutron star matter in beta equilibrium $\mu_n-\mu_p$ reaches 100 MeV at nuclear matter density, and one might expect the appearance of π^- at slightly higher densities.

One cannot, of course, neglect the interaction of the pion with the background matter. The strong repulsive s-wave pion-nucleon interaction tends to increase the pion effective mass and therefore the pion threshold density, the first point at which pions can appear. At low π^- energies the increase $\Delta E^s_{\pi^-}$ in the energy of a single π^- due to s-wave interactions with the nucleons in the matter can be estimated in terms of measured pion-nucleon scattering lengths and neutron and proton densities from

$$\Delta E^s_\pi = -\frac{2\pi\hbar^2}{m_\pi}\left[a_3 n_n + \frac{1}{3}(2a_1+a_3)n_p\right] \simeq 217(n_n-n_p)\text{MeV}, \tag{2}$$

where n_n and n_p are measured in fm^{-3}, and a_3 and a_1 are the scattering lengths in isospin 3/2 and 1/2 channels; in neutron star matter at nuclear density (2) raises the pion energy by 38 MeV. Present calculations of neutron star matter indicate that $\mu_n-\mu_p$ is always less than $m_\pi c^2+\Delta E^s_\pi$, and thus if the only pion-nucleon interaction were s-wave, pions would never appear in neutron star matter. However, Migdal,[13,14] and independently Sawyer and Scalapino,[15] pointed out that the attractive p-wave pion-nucleon interaction, which leads to the T=3/2, J=3/2 pion-nucleon scattering resonance (called Δ, N^*, or isobar), would greatly reduce the energy of pions in the medium. Such p-wave pion self-energy processes are shown in Fig. 2, where (a) corresponds to the processes $\pi^-+p\leftrightarrow n$ and $\pi^++n\leftrightarrow p$. and (b) corresponds to $\pi^-+n\leftrightarrow\Delta^-$ and $\pi^-+p\leftrightarrow\Delta^0$, as well as $\pi^++n\leftrightarrow\Delta^0$ and $\pi^++p\leftrightarrow\Delta^{++}$. The best recent calculations,[16] which take into account effects due to nuclear forces, indicate that

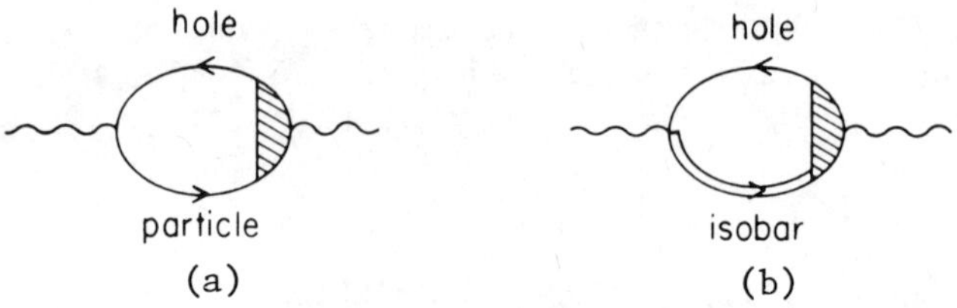

Figure 2

indeed, negative pions can appear at densities 1.5-2 times nuclear matter density.

Because pions are bosons, if they do appear in the ground state they will macroscopically occupy the lowest available mode, i.e., form a condensate as in ordinary Bose-Einstein condensation. Such a state corresponds to a classical or coherent excitation of the pion field in the medium, or alternatively, a nonvanishing ground state expectation value of the pion field; in the normal ground state, with no pions, this expectation value vanishes.

The appearance of charged pions, or more precisely, pion-like modes of excitation in the ground state, leading to a nonvanishing expectation value of the charged pion field, can actually take place through several different physical mechanisms. As Migdal has shown,[14] neutron star matter may possess a neutron hole - proton particle collective mode, or spin-isospin sound, with the quantum numbers of the π^+. Let us denote this mode, a form of zero sound, by π_s^+. Then in addition to the possibility $n \rightarrow p+\pi^-$, in which a neutron quasiparticle turns into a proton quasiparticle plus a π^-, condensation can occur via spontaneous appearance of a π^- plus the collective mode π_s^+. In addition, if the matter is initially in beta equilibrium, so that a small fraction of protons are initially present, then condensation can also occur through the mechanism $p \rightarrow n+\pi_s^+$. The existence of this additional mechanism for condensation means that matter in beta equilibrium can condense at a lower density than pure neutron matter; however studies by Weise and Brown[1] indicate that with inclusion of short ranged correlations between nucleons, matter in beta equilibrium does not condense at very different density than pure neutron matter.

The charged-pion condensed system is characterized by a non-vanishing expectation value $\langle \Pi(\mathbf{r},t) \rangle$ of the charged pion field. In the normal state this expectation value vanishes from charge conservation, as well as parity conservation, since the pion field is pseudoscalar. The π^- condensed phase is a state of broken symmetry with a complex order parameter $\langle \Pi \rangle$, and it is thus superconducting. It is useful to compare this state with a neutral Bose superfluid such as liquid He^4 and with a BCS superconductor. In He^4 the nonvanishing complex order parameter is $\langle \psi(\tilde{r},t) \rangle$, where ψ is the He^4 field operator. In a BCS superconductor one has pairing of particles of opposite spin and momenta, and $\langle \psi_\uparrow(\tilde{r},t)\psi_\downarrow(\tilde{r},t) \rangle$ is the order parameter. One can equivalently regard the π^- condensed phase as arising from a pairing of neutron particles with proton holes, and proton particles with neutron holes, with order parameter $\langle \psi_n^\dagger(\tilde{r},t)\tilde{\sigma}\psi_p(\tilde{r},t) \rangle$, the divergence of which is the nonrelativistic source of the pion field $\langle \Pi \rangle$.

From another point of view one can look at the π_s^+ mode in the normal state as the fluctuations in isospin space of the neutron isospin vectors. The neutrons normally point along the negative T_z axis; the mode π_s^+ is a small oscillation about zero of the angle α that the vector makes with respect to the negative T_z axis. At the point of condensation this mode becomes "soft" and the system beyond this prefers a nonzero value of α. The Hamiltonian is then diagonalized[15,18] by nucleon state vectors that are linear combinations of neutrons and protons.

The equilibrium conditions[19] of the π^- condensed state are first that $\langle \Pi^-(\tilde{r},t) \rangle = e^{-i\mu_\pi t} \langle \Pi^-(\tilde{r}) \rangle$, where $\mu_\pi = \mu_n - \mu_p$ is the π^- chemical potential. Even though the pions may condense into a mode $e^{i\tilde{k}\cdot\tilde{r}}$ with a nonzero momentum $\tilde{k}$, or perhaps a more complicated mode with spatially varying modulus,[20] the spatial average of the electromagnetic current must vanish in the ground state. So also must the spatial averages of the baryon current, and the electron-like and muon-like lepton currents vanish. The net charge density associated with the condensed field (in the absence of interactions dependent on $\dot{\Pi}$) is[19] $-2\mu_\pi |\langle \Pi \rangle|^2$; since generally, $\mu_\pi > 0$, this is a negative charge density. The overall system is electrically neutral.

Model calculations of the condensed state were first given by Sawyer and Scalapino.[15,20] These were improved by Baym and Flowers,[18] and Au and Baym,[21] who included "minimal effects" of nuclear forces via the Lorentz-Lorenz effect, and pi-pi interactions via a chirally symmetric phenomenological pion-nucleon Lagrangian. A simple pedagogical model calculation of the condensed phase is given in Ref. 22. There have been to date no calculations of the condensed phase including full effects of nuclear forces, isobars and pi-pi interactions, and so, a realistic assessment of the effect of condensation on the equation of state is not yet possible. A preliminary study of the effects of condensation on neutron star models has been carried out by Hartle _et al._[23]

Migdal,[14] as well as Sawyer and Yao[20] [see also Ref. 24], have considered possible π° condensation of neutron star matter. Such condensation is described by a nonvanishing real expectation value $\langle \Pi_o(\underset{\sim}{r}) \rangle$ of the neutral pion field. Because the order parameter is real the state is therefore not superconducting. Rather, since the (nonrelativistic) source of the neutral pion field is the divergence of the T=1, T_z=0 nucleon spin polarization, the neutral condensed mode $\langle \Pi_o(\underset{\sim}{r}) \rangle$ will have a finite wavevector, thus varying in magnitude in space, and leading to a spatially nonuniform spin-ordered phase, analogous to the state of electron particle-hole "pairing" produced by a Peierls' transition in a solid.

References

(1) For detailed reviews of the structure of neutron stars, see M. Ruderman, Ann. Rev. Astron. and Astrophys. 10, 427 (1972); A. G. W. Cameron, Ann. Rev. Astron. and Astrophys. 8, 179 (1970), and G. Baym and C. J. Pethick, Ann. Rev. Nucl. Sci. 25 (1975) (in press).

(2) C.-H. Yang, and J. W. Clark, Nucl. Phys. A 174, 49 (1971); N. C. Chao, J. W. Clark and C.-H. Yang, Nucl. Phys. A 179, 320 (1972); and T. Tatatsuka, Prog. Theor. Phys. 48, 1517 (1972).

(3) M. Hoffberg, A. E. Glassgold, R. W. Richardson, and M. Ruderman, Phys. Rev. Lett. 24, 775 (1970).

(4) R. Tamagaki, Prog. Theor. Phys. 44, 905 (1970); T. Tatatsuka, and R. Tamagaki, Prog. Theor. Phys. 46, 114 (1971).

(5) R. W. Richardson, Phys. Rev. D 5, 1883 (1972).

(6) T. Fujita and T. Tsuneto, Prog. Theor. Phys. 48, 766 (1972) [errata Prog. Theor. Phys. 49, 371 (1973)].

(7) K. Tkachenko, Zh. Eksp. Teor. Fiz. 50, 1573 (1966) [Sov. Phys. JETP 23, 1049 (1966)].

(8) H. P. Greenspan, The Theory of Rotating Fluids (Cambridge Univ. Press, Cambridge 1969).

(9) C.-H. Yang, and J. W. Clark, Lett. Nuovo Cim. 4, 969 (1972).

(10) G. Baym, C. Pethick and D. Pines, Nature 224, 674 (1969).

(11) G. Baym, Neutron Stars (Nordita, Copenhagen) (1970).

(12) G. Baym, H. A. Bethe and C. Pethick, Nucl. Phys. A 175, 225 (1971).

(13) A. B. Migdal, Zh. Eksp. Teor. Fiz. 61, 2209 (1971) [Sov. Phys. JETP 34, 1184 (1972)].

(14) A. B. Migdal, Phys. Rev. Lett. 31, 247 (1973).

(15) R. F. Sawyer, Phys. Rev. Lett. 29, 382 (1972); D. J. Scalapino, Phys. Rev. Lett. 29, 386 (1972); R. F. Sawyer and D. J. Scalapino, Phys. Rev. D 7, 953 (1973).

(16) S.-O. Bäckman and W. Weise, Phys. Lett. 55B, 1 (1975); G. F. Bertsch and M. B. Johnson, Phys. Rev. (to be published) (1975).

(17) W. Weise and G. E. Brown, Phys. Lett. 48B, 397 (1974).

(18) G. Baym and E. Flowers, Nucl. Phys. A 222, 29 (1974).

(19) G. Baym, Phys. Rev. Lett. 30, 1340 (1973).

(20) R. F. Sawyer and A. C. Yao, Phys. Rev. D 7, 953 (1973).

(21) C.-K. Au and G. Baym, Nucl. Phys. A 236, 500 (1974).

(22) G. Baym, D. K. Campbell, R. F. Dashen and J. T. Manassah, Phys. Lett. B (in press) (1975).

(23) J. B. Hartle, R. F. Sawyer and D. J. Scalapino, Ap. J. (in press) (1975).

(24) S. Barshay, G. Vagradov and G. E. Brown, Phys. Lett. 43B, 359 (1973).

THE HOT INTERACTING NEUTRON GAS

Jean-Robert Buchler

Department of Physics and Astronomy, University of Florida
Gainesville, Florida 32611

and

Sidney A. Coon

Institut de Physique, Université de Liège
Liège, Belgium

Abstract: We compute the interaction energy of the pure neutron gas at finite temperatures in the binary collision approximation (Brueckner-type ladder summation) for a realistic nuclear potential. The resulting equation of state is displayed for densities ranging from 10^{11} to 10^{14} g/cc and temperatures up to 10^{12} K.

The motivation for the study of the properties of the neutron gas at finite temperatures arises from astrophysics. Stellar evolutionary calculations have shown that massive stars develop small dense cores of about 1.5 solar masses which eventually become unstable when thermal decomposition of iron into alpha particles, neutrons and protons occurs. The ensuing collapse of the core proceeds essentially in free fall up to high densities when gas pressure starts to dominate again (i.e., the adiabatic index becomes sufficiently greater than 4/3) and a reverse (or bounce) of the core occurs. Two mechanisms may then be operative in the ejection of the overlying, relatively loosely bound layer of matter; a) shock ejection resulting directly from the bounce, and b) coupling between the envelope and the neutrini which are copiously emitted in the collapse. Both mechanisms have been found to be very marginal and to depend sensitively on the details of the collapse and bounce. One of the uncertainties results from the poor treatment of the equation of state, in particular the neglect of the nuclear interaction between nucleons. The bounce has been estimated to occur in the density region of 10^{12}-10^{14}g/cc and at temperatures of 2-20Mev. In this regime, the neutrons, which give the dominant contribution to the equation of state, are partially degenerate (degeneracy parameter η of order unity, i.e., $\mu \sim kT$). On the other hand, from the zero temperature results for the neutron gas we know that in this density range the nuclear interaction energy is of the same order as the kinetic energy so that its effect can hardly be neglected.

Bloch and deDominicis[1] have developed a general formalism for treating interacting fermions at an arbitrary degree of degeneracy. In the following we will apply this formalism to a novel computation of the hot interacting neutron gas. The grand potential Ω (=-pV) is expressed in terms of a linked cluster expansion similar to the Brueckner-Goldstone zero-temperature expansion.

The hamiltonian of the system is split into an unperturbed part H_o which is a sum of single particle hamiltonians h_o and a residual interaction V. The latter contains the two-body nuclear force minus whatever is already included in H_o besides the kinetic energy.

The lowest order term in the expansion is given by the well known expression

$$\Omega_o = -\frac{2}{\beta} \sum_m \ell n\, (1 + \exp(\eta - \beta e_m)) \tag{1}$$

where $\beta = \frac{1}{kT}$ is the inverse temperature in energy units, η is the degeneracy parameter, and the e_m are the eigenvalues of h_o. Simple analytic fits to $\Omega_o(\eta,\beta)$ are found in the literature, e.g. Latter.[2]

It is well known that for a strongly interacting system the perturbation expansion in terms of the bare interaction fails to converge if the latter is too strong. For a low density system, such as nucleon matter at subnuclear densities, one can recast the perturbation expansion into an expansion in terms of N-body clusters and limit oneself to the binary terms. In zero temperature many body theory the analogous treatment is known as the Brueckner K matrix approximation.

In this binary collision approximation we can write

$$\Omega = \Omega_o + \Omega' \quad , \tag{2}$$

with

$$\Omega' = \frac{1}{\beta} e^{2\eta}\, \mathrm{Tr}\, \{e^{-\beta\mathcal{H}} - e^{-\beta H_o}\}_2 \,, \tag{3}$$

where the trace goes over all antisymmetrised two particle states and where we have introduced an effective hamiltonian

$$\mathcal{H} = H_o + \tilde{V} \tag{4}$$

with the screened interaction $\tilde{V}$ defined by

$$\langle mn|\tilde{V}|rs\rangle = \langle mn|V|rs\rangle \sqrt{f_m^+ f_n^+ f_r^+ f_s^+} \ . \tag{5}$$

The Fermi functions f_k^+ (probability that state k is empty) are given by

$$f_k^+ = [1 + \exp(\eta - \beta e_k)]^{-1} \ . \tag{6}$$

Relation (3) can be cast into the form

$$\Omega' = \frac{1}{\pi} e^{2\eta} \int_{-\infty}^{+\infty} dE \ e^{-\beta E} \ \mathrm{Tr}\ \{\mathrm{atan}[\tilde{K}(E)\delta(E-H_o)]\} \tag{7}$$

where the $\tilde{K}$ operator satisfies the equation

$$\tilde{K}(z) = \tilde{V} + \tilde{V} \frac{pp}{z-H_o} \tilde{K}(z) \tag{8}$$

or the corresponding unscreened [see definition (5)] matrix K satisfies the integral equation

$$\langle mn|K(z)|pq\rangle = \langle mn|V|pq\rangle + \sum_{rs} \langle mn|V|rs\rangle \frac{f_r^+ f_s^+}{z-e_r-e_s} \langle rs|K(z)|pq\rangle . \tag{9}$$

Except for the form of the exclusion function $f_r^+ f_s^+$, this equation is identical with Brueckner's K matrix equation, (Ref. 3).

Since the K matrix is diagonal in total momentum this six-dimensional equation can be reduced to a three-dimensional one. In order to be able to perform a partial wave expansion we take an average over the angle between total and relative momenta P and k, respectively, i.e. we replace $f_r^+ f_s^+$ by

$$Q(k,P;\eta,\beta) \equiv \frac{1}{2} \int_{-1}^{+1} f^+(\vec{k} + \frac{\vec{P}}{2}) \ f^+(\vec{k} - \frac{\vec{P}}{2}) \ d(\hat{k}\cdot\hat{P})$$

$$= \frac{1}{1-b^2} \left(1 - \frac{1}{\gamma Pk} \ \ell n \left(\frac{1+a^-}{1+a^+}\right)\right)$$

where

$$\gamma = \beta\hbar^2/2M$$

$$a^{\pm} = \exp[\eta - \gamma(k \pm \frac{P}{2})^2]$$

$$b = \exp[\eta - \gamma(k^2 + \frac{P^2}{4})] \quad .$$

Such an angle average is known to give good results in the zero temperature conditions.

The resultant integral equations are now one dimensional, but coupled because of the tensor force (K is non-diagonal in orbital angular momentum):

$$K^{JS}_{LL'}(k,k',P,z;\eta,\beta) = V^{JS}_{LL'}(k,k')$$

$$+ \frac{2}{\pi}\sum_{\ell}\int_0^{\infty} dk'' V^{JS}_{L\ell}(k,k') \frac{Q(k',P;\eta,\beta)}{z - \frac{\hbar^2 k''^2}{M}} K^{JS}_{\ell L'}(k'',k',P,z;\eta,\beta) \; .$$

The potential $V^{JS}_{LL'}(k,k')$ is the momentum transform of the nuclear potential for which we have chosen the isospin triplet Reid potential[4]

$$V^{JS}_{LL'}(k,k') = C^{JS}_{LL'}\int_0^{\infty} dr\, r^2 j_L(kr) j_{L'}(k'r) V^{JS}_{LL'}(r) \quad ,$$

where the constants $C^{JS}_{LL'}$ absorb the angular and spin factors. For the numerical solution we have followed essentially the method of Haftel and Tabakin.[5]

In order to compute the trace we have to diagonalize K with the result,

$$\Omega' = \sum_{mn} e^{2\eta-\beta(e_m+e_n)}\; 2\,\Sigma(2J+1)\delta_{Jimn}(\eta,\beta)$$

$$= -\frac{2}{\pi^3}\frac{\hbar^2}{M} e^{2\eta}\int_0^{\infty} dP\, P^2 e^{-\beta\frac{\hbar^2 P^2}{4M}} \int_0^{\infty} dk\, k e^{-\beta\frac{\hbar^2 k^2}{M}} \sum_{J,i}(2J+1)\delta_{Ji}(k,P;\eta,\beta)$$

where the $\delta_{Ji} = \operatorname{atan}[Q(k,P;\eta,\beta)K_{Ji}(k,k,P,z=\frac{\hbar^2 k^2}{M};\eta,\beta)]$ are the eigenphases of $\tilde{K}$; J is the total angular momentum and i denotes the various S and L combinations allowed by the symmetry of the particles. In the limit of low temperatures $T \to 0$, $\eta/\beta \to \varepsilon_F$, one obtains

$$\Omega' \simeq \Sigma(1 - f^{+}_{m})(1 - f^{+}_{n})\langle mn|K|\, mn\rangle$$

which is numerically more convenient and accurate when the gas becomes sufficiently degenerate.

Once the grand potential Ω is known, all other thermodynamic quantities can be derived from it, in particular the density

$$n = -\beta\left(\frac{\partial\Omega}{\partial\eta}\right)_\beta = n_0 + n' .$$

Analytic fits are available[2] for the unperturbed density n_0

$$n_0 = -\beta\left(\frac{\partial\Omega_0}{\partial\eta}\right)_\beta = 2\sum_m (1 - f_m^+) .$$

The perturbation correction n' to the density is obtained by numberical differentiation with respect to Ω'.

In our computations we have used all orbital angular momentum states L with $J \leq 2$ (except $^3F_2$). At low temperature ($\beta \gtrsim 10\text{MeV}^{-1}$), however, the $^1S_0$ state was so dominant that all others would be left out with negligible error.

The resulting equation of state is shown in the figure in solid lines. For comparison, the unperturbed pressure is shown in dashed lines. The curves are labelled according to the inverse temperature (in MeV). The one labelled $\beta = \infty$ corresponds to the pressure obtained with a Brueckner code (described in Ref. 6); for comparison with the finite temperature results, no single particle (or hole)

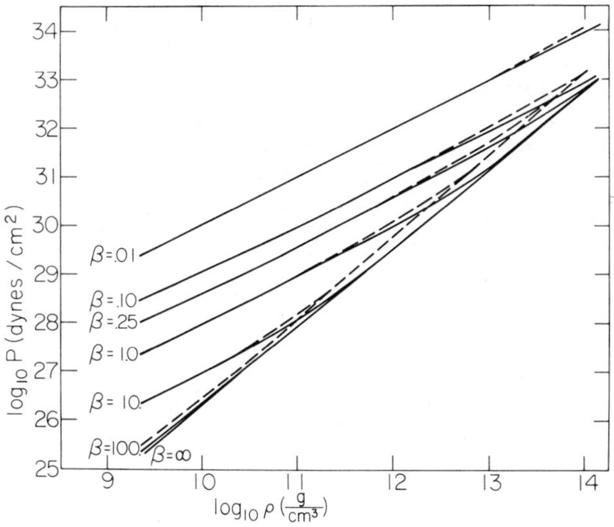

Figure 1

potential was used however. The confluence of the finite β curves provide an excellent check on our code since the two approaches, despite their similarity, derive from basically different perturbation expansions. The Brueckner-Goldstone expansion is done at constant density with perturbative corrections to the chemical potential (the Hugenholtz-van Hove theorem, $\mu = \frac{d\varepsilon}{d\zeta}$, is generally only satisfied if all orders are considered), whereas the Bloch expansion is done at constant chemical potential. In the language of perturbation theory our expression for the second virial coefficient sums a wider class of diagrams than just the Brueckner ladder graphs with intermediate particle states. It is also more general than the high temperature Beth-Uhlenbeck[7] formula in that intermediate scatterings into occupied states are prevented. In the high density (close to nuclear), low temperature regime, however, we would expect our approach to be only approximate as we have not renormalized our single particle energies by the inclusion of a one body (self-consistent or other) potential in H_o. We have therefore not pushed our computations beyond 10^{14}g/cm^3 where this latter effect becomes important. The inclusion of the Pauli operator in intermediate states, however, makes our treatment more appropriate than the usual zero temperature phase-shift approximation (Brueckner et al., Ref. 8). An inspection of the graph shows that, as expected, the neutron interaction energy starts to become important right in the region where the core-bounce occurs. Its effect should be measured not so much by the mere reduction of the pressure at a given density, as by the reduction in the stiffness of the pressure curve (adiabatic index = $\frac{d\ln p}{d\ln \zeta}$) which governs the bounce point.

Although the neutrons are the dominant component of the stellar matter in consideration there can be of the order of 10 percent or more protons. The effect of the latter on the interaction pressure is expected to outweigh their numerical importance primarily because of the additional isospin singlet interaction (as a reminder, I mention that nuclear matter is bound but the pure neutron gas is unbound at all densities). The present formalism can readily be extended to include the effects of a finite proton concentration at the expense of making Ω depend now on both neutron and proton degeneracies.

Finally, to get the actual equation of state required in the hydrodynamic core evolution, the contribution of electrons and possibly muons will be included. When this additional study is completed, one hopes to have eliminated one of the uncertainties which plague the study of supernovae and neutron star formation.

References

(1) C. Bloch, Studies in Statistical Mechanics, Vol. III, ed. de Boer and Uhlenbeck (North-Holland, 1965) referred to as BD; for further details see C. Bloch, and C. de Dominicis, Nucl. Phys.

7 459 (1958); 10 181 (1959); 10 509 (1959); J. M. Luttinger and J. C. Ward, Phys. Rev. 118 1417 (1960); W. Kohn and J. M. Luttinger, Phys. Rev. 118 41 (1960).

(2) R. Latter, Phys. Rev. 99 1854 (1955).

(3) K. A. Brueckner, The Many-Body Problem, Les Houches, 1958, (Methuen, Wiley, Dunod, 1959) or B. D. Day, Rev. Mod. Phys. 39 719 (1967) or H. A. Bethe, Ann. Rev. Nucl. Sci. 21 93 (1971).

(4) R. V. Reid, Ann. Phys. (N.Y.) 50 411 (1968).

(5) M. I. Haftel and F. Tabakin, Nucl. Phys. A 158 (1970).

(6) K. A. Brueckner, S. A. Coon and J. Dabrowski, Phys. Rev. 168 1184 (1968).

(7) E. Beth and G. E. Uhlenbeck, Physica 4 915 (1937).

(8) K. A. Brueckner, C. A. Levinson, and H. M. Mahmoud, Phys. Rev. 95 217 (1954); K. A. Brueckner, Phys. Rev. 96 508 (1954).

A GENERAL THEORY FOR SUPER PHENOMENA IN PHYSICS

A. J. Coleman

Department of Mathematics, Queen's University

Kingston, Ontario, Canada

Abstract: Super phenomena in physics are explained as resulting from a large eigenvalue of the second-order reduced density of the system and as always occurring, for sufficiently low temperatures, if the corresponding second-order reduced hamiltonian possesses one or more bound states. The excitation spectrum of He^3 and of He^4 is interpreted in terms of pair transitions.

I. INTRODUCTION

The recent paper[1] of H. W. Jackson concludes that the data on neutron scattering by He II is consistent with zero condensation in the one-particle density matrix. Certainly, if any such condensation obtains, it is much too small, 2 percent at most, to allow one to interpret the superfluid "component" of He II as consisting of the atoms in the lowest one-particle state. Superfluidity is essentially a cooperative phenomenon. Interaction between helium atoms must be crucial. We are, therefore, forced to recognize that superfluidity should manifest itself in the second or higher order density matrix. This is a view which the present author advanced several years ago.[2-10] In addition to the arguments of Jackson, further evidence is provided by the fact that the phase diagrams for the He^3 and He^4 are topologically equivalent. The type A region of He^3, which has properties suggestive of a liquid crystal, would correspond to the b.c.c. region in solid He^4 at the top of the lambda curve.

The present paper brings together the arguments scattered

through the references to the effect (i) that transitions to superfluidity, to superconductivity and to magnetic ordering are associated with a large eigenvalue of the two-matrix; (ii) that the "phonons" and "rotons" of Landau's and Feynman's treatment of He are picturesque descriptions for pair excitations; (iii) that superconductivity and magnetic ordering occur when the second-order reduced hamiltonian has a bound state; (iv) that (a) superconductivity, (b) magnetic ordering or (c) both, occurs according as the large eigenvalue of the two matrix corresponds to (a) a singlet, or (b) a triplet state of the two-matrix or (c), both.

II. PAIRON ANALYSIS OF ENERGY

The statistical mechanics of free particles is relatively simple chiefly because the total energy of an N-particle system can be expressed as a sum of the energies E_i of one-particle states weighted by one-particle occupation numbers: $[\exp \beta(E_i - \mu)\ 1]^{-1}$. By treating quasi-particles rather than particles, Landau was able to extend this approach to the low-lying excitations of real systems. However, when interaction between particles or quasi-particles is a dominant feature, as in most physical systems, this simple approach is inappropriate.

For a system of indistinguishable particles with hamiltonian of the form

$$H = \Sigma H(i) + \sum_{i<j} H(i,j)$$

it is well-known[8] that the exact energy is given by

$$E = \binom{N}{2} \text{ tr } KD = \binom{N}{2} \Sigma p_i \epsilon_i \qquad (1)$$

where the reduced hamiltonian

$$K = (N-1)^{-1}[H(1)+H(2)] + H(12) \quad , \qquad (2)$$

the 2-matrix or second-order reduced density matrix,

$$D^2(12,1'2') = \int_{3\ldots N} \psi(123\ldots N)\bar{\psi}(1'2'3\ldots N) \quad , \qquad (3)$$

$$K\gamma_i(12) = \epsilon_i \gamma_i(12) \quad , \qquad (4)$$

and

$$p_i = \langle \gamma_i | D^2 \gamma_i \rangle \quad . \qquad (5)$$

The eigenfunctions γ_i, and D^2 are normalized so that

$$\Sigma p_i = 1 \tag{6}$$

We refer to a two-particle function as a geminal and to the occupant of a geminal as a pairon. In analogy with the one-particle picture, we are constrained to interpret equation (1) by saying that $\binom{N}{2}p_i$ is the number of pairons occupying the geminal γ_i of energy ϵ_i. Thus the exact energy of our interacting system is a weighted sum of the eigenvalues of K. However unlike the situation for free fermions and bosons, there is no known simple formula for the weights p_i.

When the spectrum of K is not discrete (1) may be replaced by the integral form

$$E = \binom{N}{2} \int \epsilon \, p(\epsilon) \, d\epsilon \tag{7}$$

where $p(\epsilon)$ is the pairon occupation density, and we include the discrete and continuous portions of the spectrum of K by employing the Stieltjes-Lebesque integral. If $q(\epsilon)$ is the density of states for K, then

$$f(\epsilon) = p(\epsilon)/q(\epsilon) \tag{8}$$

should be interpreted as the pairon probability distribution. When the individual particles are fermions, it is known [4, eq. (12)] that

$$0 \leq f(\epsilon) \leq (N-1)^{-1} \tag{9}$$

If pairons were bosons or fermions, the upper bound in (9) would be replaced by 1 and $\binom{N}{2}^{-1}$ respectively. Thus a pair of fermions can be regarded neither as a boson nor a fermion. Indeed, because of the complex interrelations among the p_i demanded by N-representability, it appears unlikely that the statistics of fermi pairs is given, even approximately, by Gentile's intermediate statistics. These issues are discussed more fully in Ref. 9.

III. SUPER TRANSITIONS

Inspired by C. N. Yang[11] but using a simpler argument, we pointed out[3,4] that infinite long-range order may occur in a macroscopic system of N identical particles if an eigenvalue of the 2-matrix is of order N^{-1}. By this, we mean that in the thermodynamic limit the eigenvalue remains greater than αN^{-1} for some fixed $\alpha > 0$. Since $p_i \leq \lambda_1^2$, where λ_1^2 is the largest eigenvalue of D^2, it is only in this situation that p_i of (5) can attain order βN^{-1}. But if it does then, according to (1) there is an extensive energy $\frac{1}{2}\beta(N-1)\epsilon_i$

associated with the geminal γ_i. This would be manifest as infinite range order since a two-particle function describes a correlation.

In the case of a system of fermions, it is known[1] that the eigenvalues of the 2-matrix, $\lambda_i^2 \leq (N-1)^{-1}$ and that this upper-bound can be attained arbitrarily closely when the wave function of the system is a so-called antisymmetrized geminal power (AGP). This type of function is also called a pairing or projected BCS function. The ansatz about the wave function is[7] the cornerstone of the BCS model for superconductivity.

For a system of bosons, the statistics, as such, places no limitation on λ_i^2 since it is trivial to display a wave function, corresponding to non-interacting particles, for which $\lambda_1^2 = 1$, $\lambda_i^2 = 0$ for $i > 1$. However, the hard-core which seems to be a universal characteristic of the potential between identical physical particles effectively prevents total condensation of boson pairs in one geminal. We shall argue in section V that for He^4 an upper bound of $8N^{-1}$ on λ_i^2 is imposed by the potential. Thus, $f(\epsilon)$ of (8) is very largely determined by the dynamics of the system so it is too much to hope that f is a universal function as occurs[9] for non-interacting particles.

IV. SUPERCONDUCTIVITY AND MAGNETIC ORDERING

Froehlich[12] asserted that the first observations of the isotope effect for which $\alpha = 0.5$ in $T_c = m^{-\alpha}$, "proved" that superconductivity was due to an interaction between electrons mediated by the phononfield of the lattice vibrations. However, for zirconium there is[13] no isotope effect (α=0) and for uranium it is negative (α=-2). This would equally well "prove" that superconductivity was not due to a phonon mediated interaction. It seems to the author that the theory of superconductivity has been distorted by the habit of physicists of taking the independent-particle picture as their starting point and by their belief that superconductivity is an esoteric effect which requires an esoteric explanation! As Matthias and others have shown most convincingly, superconductivity or magnetic ordering is a universal phenomenon at sufficiently low temperatures. At Sanibel, 1975, Matthias even predicted that pure gold will prove to be a superconductor at sufficiently low temperatures. The fact that so-called "Cooper pairs" are possible is due in the first instance to the fact that the positive charge on the nuclei of the lattice hold the electrons within talking distance of each other. This truth is just too obvious to have been noticed by the theoretical physicists who discussed superconductivity and have been pursuing some more esoteric explanation.

For a neutral lattice in the adiabatic approximation, the

reduced Hamiltonian is given[7] by

$$K = K_1 + K_2(T) + L \tag{10}$$

where

$$(N-1)K_1 = -\frac{1}{2}\Delta_1 - \frac{1}{2}\Delta_2 - \Sigma_{\kappa} Z_{\kappa}(|\underline{r}_1-\underline{\kappa}|^{-1} + |\underline{r}_2-\underline{\kappa}|^{-1})$$

$$+ (N-1)\,|\underline{r}_{12}|^{-1} \quad . \tag{11}$$

L is the contribution from the Coulomb interaction and vibrational energy of the ions averaged over the ionic wave function. $K_2(T)$ is the sum of (i) a one-body potential arising from the part of the Coulomb interaction between the ions and electrons not included in K_1 and due to the fact that the ions are not fixed, and (ii) one and two body terms involving spin and phonons.

Obviously, the nature of the spectrum of the operator K is of paramount importance for solid-state physics. A theorem of Zhislin[14] asserts that if $\Sigma Z_{\kappa} > N-1$, as is the case for a neutral lattice, then K_1 will have an infinity of bound states with eigenvalues below the lower edge of the continuous spectrum. However, if length is rescaled by N and the lattice is allowed to collapse so that all $\underline{\kappa} = 0$, K_1 becomes the hamiltonian for H^- which is known to have only one eigenvalue appreciably below the edge of the continuous spectrum.

Thus while Zhislin's theorem assures us that K_1 has a large supply of bound states, the behaviour of the limiting case H^- suggests that there will be a significant gap between the ground state and the first excited state of K_1. It is our contention that the onset of superphenomena coincides with a condensation of pairons in the lowest energy level - or levels - of K_1 such that p_1 is of order N^{-1}. However, at temperatures $T > T_c$, the spectrum of K is not completely dominated by K_1. Apparently, for $T > T_c$, the spin or phonon terms in $K_2(T)$ perturb the K_1 spectrum in such a manner as to mask the energy gap between the ground state and first excited levels of K_1. Certainly, there is a sense in which K_1 is the dominant part of the expression (10) for K. The contribution L is a very large energy tending to blow the lattice, or liquid apart. The cohesion of lattice or liquid is possible only because the negative binding energy associated with K_1 dominates the large repulsive forces associated with L. The energy $K_2(T)$ is very much smaller than either K_1 or L. But since the latter two cancel one another, $K_2(T)$ in fact at certain points controls the "balance of power". Depending on the nature of the particular system in question, the most important contribution to $K_2(T)$ could come from either phonon or spin terms, so the present theory provides a framework which can readily accommodate the fact emphasized by Matthias that depending on the column of Periodic Table, T_c depends more or less on phonon or spin effects.

Presumably, if the ground state of K is a triplet, condensation of pairons in it would be manifest as magnetic ordering. Hopefully, the present theory will provide insight into the correlation, upon which Matthias commented in his McGill lectures[13], between superconductive and metallurgical properties of metals.

It is probably unreasonable to expect that the complex patterns observed during the gradual onset of superconductivity through a bulk specimen can be adequately described by one geminal. The low-lying levels of K might be in a dense narrow band which would encourage the appearance of many "large" eigenvalues in D^2. Suppose[5] that in the thermodynamic limit $ND^2 \to D_c^2$ where $tr(D_c^2) = \alpha > 0$. Then D_c^2 governs the correlation in the system and generalizes the Ginzburg-Landau order parameter.

V. SUPERFLUIDITY

In his comprehensive book on helium, Wilks[15] remarks at several points that the behaviour of the two isotopes is not greatly different. The chief differences are a consequence of their unequal masses and of the nuclear spin of He^3. When Wilks wrote, the one exception he could point to was the extraordinary transition of He^4 to the superfluid state and the absence of any analogous observed transition in He^3.

However, we now know that He^3 makes a similar transition to a superfluid state at about 2 millidegrees. The phase diagram is remarkably similar to that of He^4 if we correlate the superfluid liquid phase of He^4 with the B phase of He^3 and the solid b.c.c. phase of He^4 with the A phase of He^3 which has been described as showing properties similar to a "liquid crystal". He^3, of course, has peculiar magnetic properties resulting from its nuclear spin. The difference in statistics seems to have forced the superfluid transition in He^3 to a much lower temperature than in He^4 but the general nature of the transition in the two cases seems almost identical. One, therefore, expects that they are both explicable by the same mechanism. Since He^3 is a fermion, its transition is not susceptible of explanation as a bose condensation accompanied by a large eigenvalue of the 1-matrix. Jackson's paper[1] frees us from any need to subscribe to the current orthodoxy which does attempt such an explanation of the He^4 transition.

It seems much more reasonable to the present author to explain both superfluid transitions as resulting from a large eigenvalue of D^2 which for He^3 cannot be larger than N^{-1} because of the statistics, whereas for He^4 it could be of order βN^{-1} where β is a packing factor which is about 8 for nearest neighbours.

In first approximation, we treat helium atoms as point particles interacting by means of a potential $V(\rho)$ where ρ is the distance between two atoms. Then the reduced hamiltonian

$$K = (N-1)^{-1}[-\frac{1}{2}\Delta_1 - \frac{1}{2}\Delta_2] + V(\rho) \quad (12)$$

is separable when referred to centre of mass or extracule coordinates

$$R = \frac{1}{2}(\underline{r}_1 + \underline{r}_2) \quad (13)$$

and relative or intracule coordinates

$$\underline{\rho} = \underline{r}_2 - \underline{r}_1 \quad . \quad (14)$$

However, if we consider a sphere of helium of radius a and seek solutions of (4) which vanish for $r_i \geq a$, the boundary conditions are not separable. However, in 1964 Dr. Bryce McLoed of Wadham College, Oxford pointed out to the author that the region in 6-space,

$$\Sigma : r_1 \leq a \quad , \quad r_2 \leq a \quad (15)$$

contains

$$\Sigma' : R \leq \frac{1}{2} a \quad , \rho \leq a \quad (16)$$

and is contained in

$$\Sigma'' : R \leq a \quad , \rho \leq 2a \quad . \quad (17)$$

That is

$$\Sigma' \subset \Sigma \subset \Sigma'' \quad . \quad (18)$$

Further, the eigenvalue problem (4) for the regions Σ' and Σ'' are separable. Since making boundary conditions less or more stringent respectively raises or lowers an eigenvalue, it follows that the n-th eigenvalue of (4) on region Σ is bracketed by the n-th eigenvalues of the two separable problems of (4) on Σ'' and Σ'.

Extracules carry momentum and energy and behave like free particles. The intracules have zero momentum but are ascribed a mass and carry vibrational and rotational energy. The extracules and intracules will therefore play key roles in the discussion of transport and condensation phenomena respectively. In particular, density and long-range correlation will be determined by the intracules.

According to the well-known formula of H. Weyl, the number of linearly independent eigenfunctions of (12) with eigenvalue less than or equal to C is given by

$$N(C) = \frac{(N-1)^3}{48\pi^3} \int (C-V)^3 \, d^3\underline{r}_1 \, d^3\underline{r}_2 \tag{19}$$

where the integral is taken over the region in Σ for which $C - V > 0$. Since the interaction potential $V(\rho)$ between two helium atoms varies as ρ^{-6} at large distances, for $C > 0$ we obtain a good approximation to the integral (19) by setting $V = 0$, so

$$N(C) = \frac{v^2(N-1)^3}{48\pi^3} C^3 + N(0) \quad , \tag{20}$$

where v is the volume of the specimen of helium. Thus if $dN(C) = q(C)\, dC$, the density of states for $C > 0$ satisfies

$$q(C) = A\, C^2 \quad , \tag{21}$$

where A is a constant.

The expression in (21) for the density of pairon states will be exact if $V = 0$. This is the case $s = 2$ discussed on p. 314 of Landsberg[16] who shows that for free bosons with density of states as in (21), the specific heat varies as T^3 for low temperature. It was this observed property of He^4 which motivated Landau to introduce the concept of phonons. Further, there will be a discontinuity in the specific heat which, while it is not lagarithmic, approximates very closely to the lambda transition.

The factor (N-1) in (2) has the effect of changing the very shallow helium potential well into one that is quite deep, so that there are many bound states for the intracule. However, not all pairons can condense in the lowest intracule because this would imply that all the helium atoms were within roughly 5Å of each other. On the average, they are at a distance of about half the diameter of the specimen. The hard-core of the potential prevents closer approach than about 2.5 Å, so it is dynamics rather than statistics which prevents B.E. condensation of helium pairons. In fact, the lowest intracule level will characterize nearest neighbours and it is known experimentally that for temperatures between 1.6 K and 5.5 K the number of nearest neighbours lies between 7.0 and 8.4. For a body-centred cubic lattice, it would be exactly 8. The number of nearest-neighbour pairs is $8N/2 = 4N$ which suggests that $4N/\binom{N}{2} = 8N^{-1}$ is an approximation to the upper bound for occupancy of the lowest intracule state. When it is maximally occupied every atom is locked to all its nearest neighbours by a pairon state with zero rotation, but of which the extracule could have non-zero momentum giving rise to a highly correlated flow. Excited intracule states could have rotation and thus explain the elementary vortices.

We therefore propose that the lambda transition for He^4 be explained as occurring when occupancy of the pairon states associated with the lowest intracule level rises to order N^{-1}. Exactly the same mechanism explains the superfluid transition in He^3. However, in the latter case the onset of the transition is inhibited by the dynamics and by the statistics of the fluid. Whereas the pairon states of He^4 corresponding to nearest neighbours can have occupancy $8N^{-1}$, to second neighbours, $64N^{-1}$ and so, for He^3 none of these can exceed N^{-1}.

The neutron scattering experiment on helium involve transfer of momentum and will therefore be explicable only in terms of excitation of extracules. Possibly, the main features of the observed dispersion curve can be explained in terms of single pairon excitations. For example, the part of the dispersion curve between change in momentum of 1.9 and 2.4 $\text{\AA}^{-1}$ is fairly well represented by a free extracule.

VI. CONCLUSION

We do not pretend that the present paper gives a complete or quantitatively precise theory. However, we do claim that it sets forth a simple and unified set of ideas in terms of which the widespread superphenomena in physics may be understood. Unlike most other such attempts which take an independent-particle model as their starting point, we insist that correlation, and therefore, interaction is of the essence of superphenomena. For this reason, there is no alternative but to begin by looking at the second order matrix.

References

(1) H. W. Jackson, Phys. Rev. A14, 278 (1974).

(2) A. J. Coleman, Rev. Mod. Phys. 35, 668 (1963).

(3) A. J. Coleman, J. Math. Phys. 6, 1425 (1965).

(4) A. J. Coleman, Can. J. Phys. 42, 226 (1964).

(5) A. J. Coleman, Can. J. Phys. 45, 1271 (1967).

(6) A. J. Coleman, Phys. Rev. Lett. 13, 406 (1964).

(7) A. J. Coleman, Functions of BCS Type and Superconductivity, Preprint No. 103 - Quantum Chem. Group, Uppsala University, 1963 (unpublished).

(8) A. J. Coleman, Int. J. Quantum Chem. 1, 457 (1967).

(9) A. J. Coleman, Density Matrix Methods in Quantum Chem., Notes by J. Veverka based on a set of lectures at Sanibel, 1965, (unpublished).

(10) A. J. Coleman and S. Pruski, Can. J. Phys. 43, 2142 (1965).

(11) C. N. Yang, Rev. Mod. Phys. 34, 694 (1962).

(12) H. Froehlich, Rep. Prog. Phys., 24, 1 (1961).

(13) Matthias, B. T., Systematics of Superconductivity, in Superconductivity, Vol. I, Summer Study Institute, McGill University, (Gordon and Breach, N.Y., N.Y.), 1969.

(14) G. M. Zhislin, Trudi. Mosk. Math. Obsc. 9, 81 (1960).

(15) J. Wilks, Liquid and Solid Helium, (Oxford Univ. Press), 1967.

(16) P. T. Landsberg, Thermodynamics, (Interscience Publishers, N.Y., N.Y.), 1967.

C-NUMBER REPRESENTATION FOR MULTILEVEL SYSTEMS AND THE QUANTUM-CLASSICAL CORRESPONDENCE

R. Gilmore[*], Charles M. Bowden[†], and Lorenzo M. Narducci[§]

[*]University of South Florida, Tampa, Florida 33600

[†]Physical Sciences Directorate, Redstone Arsenal, Alabama 35809

[§]Worcester Polytechnic Institute, Worcester, Massachusetts 01609

Abstract: We use the r-mode direct product coherent states as generating functions for r-level coherent states which are ideal for the description of collective behavior of an ensemble of N identical r-level atoms or molecules, where the r-levels are not necessarily evenly spaced. It is noted that the Lie algebra for an r-level system can be given a realization in terms of bilinear combinations of boson creation and annihilation operators. This provides a homomorphism from the algebra describing a multimode system to the algebra describing a multilevel system. This in turn provides a homomorphism from the multimode coherent states and their diagonal projectors onto the multilevel coherent states and their diagonal projectors. The action of a creation or annihilation operator on a multimode projector can be replaced by the action of a first order differential operator. This differential realization of operators is called a $\mathcal{D}$-algebra. The $\mathcal{D}$-algebras for multimode systems are constructed, and the homomorphisms from multimode to multilevel systems are used to project the multimode $\mathcal{D}$-algebras onto the multilevel $\mathcal{D}$-algebras. We illustrate the utility of the multilevel $\mathcal{D}$-algebra by specializing to systems of N, identical two-level atoms, and use the coherent states together with the $\mathcal{D}$-algebra to derive a c-number, phase space description of the atomic density operator equation of motion. Further application is shown by the use of the $\mathcal{D}$-algebra to represent the equation of motion for the diagonal element of the density operator in a coherent state representation as a partial differential equation. The coherent state representation is used to show the formal equivalence between these two c-number differential equations. Further, it is shown that their solutions are related by a convolution.

I. INTRODUCTION

The construction of c-number differential algebras, which we shall call $\mathcal{D}$-algebras, has been motivated by a desire to study in detail the dynamical properties of quantum mechanical systems. The equation of motion for an operator, in particular the density operator $\rho(t)$, involves a commutator. It is the presence of the commutator together with the fact that the system operators do not generally commute that makes the equation of motion particularly difficult to solve. While it is often possible to solve for the diagonal matrix elements of $\rho(t)$, it is generally not possible to determine the off-diagonal matrix elements without invoking additional assumptions. Such assumptions always diminish our knowledge of quantum correlation effects which can be very important in the description of cooperative phenomena.

It is therefore desirable to look for a representation in which every physically reasonable operator can be expressed in diagonal form without any loss of information. This can only be done using a nondenumerable overcomplete set of "continuous basis" states. In this representation, the equation of motion of the density operator can be mapped exactly into a classical-like c-number equation of motion in a phase space defined by the variables of the coherent state parameterization. The construction of this c-number equation is greatly facilitated by the development of a $\mathcal{D}$-operator algebra. This constitutes an assignment of a first-order linear differential operator to each operator in the (Lie) algebra characterizing the system under consideration.

In the following section, we develop the $\mathcal{D}$-algebra for a single mode of the electromagnetic field, for r-independent modes of the electromagnetic field (multimode system) and for N identical r-level atoms (multilevel system).

In section III, we specialize the $\mathcal{D}$-algebra to application to an ensemble of N identical two-level atoms. We use the coherent states as a basis, together with the $\mathcal{D}$-algebra to map a class of master equations into c-number form. In this section, we also use the $\mathcal{D}$-algebra to transform the equations of motion for the diagonal elements of the density operator into partial differential equations. We show the formal equivalence between these separate equations by connecting the respective solutions by a convolution integral. It is shown that complete statistical information is contained in the solution to either differential equation with the appropriate initial conditions.

II. THE $\mathscr{D}$-ALGEBRA

A. Single Mode $\mathscr{D}$-Algebra

A single mode of the electromagnetic field is characterized by the creation and annihilation operators $a^\dagger$ and a, by the number operator $a^\dagger a$, and by the identity I, which obey the usual boson commutation relations,

$$\begin{aligned} [a^\dagger a, a^\dagger] &= a^\dagger \\ [a^\dagger a, a] &= -a \\ [a, a^\dagger] &= I \\ [\cdot, I] &= 0 \quad . \end{aligned} \tag{2.1}$$

Basis vectors in this Hilbert space are often chosen to be eigenstates of the number operator (Fock states)

$$a^\dagger a|n\rangle = n|n\rangle \quad , \quad n = 0,1,\ldots \tag{2.2}$$

In this representation an arbitrary operator $\hat{F}$ is generally not diagonal, i.e.,

$$\hat{F}(t) = \sum_{m,n} |m\rangle\, F_{mn}(t) \langle n| \qquad F_{mn}(t) \neq 0 \quad m \neq n \quad . \tag{2.3}$$

A "diagonal" representation for most physically reasonable operators can be constructed by introducing the field-coherent states[1] $|\alpha\rangle$ (α complex), obtained by applying a unitary transformation to the ground state $|0\rangle$,

$$|\alpha\rangle = e^{\alpha a^\dagger - \alpha^* a}|0\rangle \tag{2.4}$$

$$= \sum_{m=0}^{\infty} |m\rangle\, f_m(\alpha) \tag{2.5}$$

where

$$f_m(\alpha) = e^{-\alpha^*\alpha/2} \frac{(\alpha)^m}{\sqrt{m!}} \quad . \tag{2.6}$$

These coherent states have a one-to-one correspondence with points α in the complex plane, and form an overcomplete set of basis elements in the Hilbert space. It is therefore possible to expand any operator in terms of these coherent states:

$$\hat{F}(t) = \int \frac{d^2\alpha}{\pi} \int \frac{d^2\beta}{\pi} |\alpha\rangle \; F(\alpha^*,\beta,t) \langle\beta| \, . \tag{2.7}$$

The operator $\hat{F}$ is then said to be represented by the function F. The overcompleteness of the states $|\alpha\rangle$ renders such representations nonunique. In fact, the nonuniqueness may be exploited to construct a <u>diagonal</u> representation for almost any physically reasonable operator.[2]

$$\hat{F}(t) = \int \frac{d^2\alpha}{\pi} |\alpha\rangle F(\alpha,t) \langle\alpha| \quad . \tag{2.8}$$

It is often necessary to consider the action of an operator $\hat{A}$ (i.e., $a, a^\dagger$) on F from either the left or the right. Thus $\hat{A}$ acts on the "projectors" $|\alpha\rangle\langle\alpha|$ in eq. (2.8) as follows:

$$\hat{A}|\alpha\rangle\langle\alpha| = \hat{A} \sum_{m,n} |m\rangle \frac{e^{-\alpha^*\alpha} \alpha^m (\alpha^*)^n}{\sqrt{m!n!}} \langle n| \tag{2.9}$$

$$= \sum_{m,n} \frac{e^{-\alpha^*\alpha} \alpha^m (\alpha^*)^n}{\sqrt{m!n!}} \hat{A}|m\rangle \langle n| \quad . \tag{2.10}$$

It is always possible to construct a c-number differential operator $\mathcal{D}^L(\hat{A})$ having the same effect on the parameters α and α^* of $f_m(\alpha)$ $f_n^*(\alpha)$ as does the operator $\hat{A}$ acting on the projector $|\alpha\rangle\langle\alpha|$ from the left,

$$\hat{A}|\alpha\rangle\langle\alpha| = \sum_{m,n} |m\rangle \, \mathcal{D}^L(\hat{A}) \left\{ \frac{e^{-\alpha^*\alpha} \alpha^m (\alpha^*)^n}{\sqrt{m!n!}} \right\} \langle n| \quad . \tag{2.11}$$

In particular

$$\mathcal{D}^L(a) = \alpha$$

$$\mathcal{D}^L(a^\dagger) = \frac{\partial}{\partial\alpha} + \alpha^* \quad . \tag{2.12}$$

Thus, we have an algebra which we call a $\mathcal{D}$-algebra, and the following properties are easily verified:

$$\mathcal{D}^L(r\hat{A} + s\hat{B}) = r\mathcal{D}^L(\hat{A}) + s\mathcal{D}^L(\hat{B}) \tag{2.13a}$$

$$\mathcal{D}^L(\hat{A}\hat{B}) = \mathcal{D}^L(\hat{B}) \; \mathcal{D}^L(\hat{A}) \tag{2.13b}$$

$$\mathcal{D}^L([\hat{A},\hat{B}]) = [\mathcal{D}^L(\hat{B}), \mathcal{D}^L(\hat{A})] \quad . \tag{2.13c}$$

Similar differential operator realizations for arbitrary operators $\hat{A}$ acting from the right also exist. These are related simply to the left realizations as follows:

$$\mathcal{D}^R(\hat{A}) = \left(\mathcal{D}^L(\hat{A}^\dagger)\right)^* \quad . \tag{2.14}$$

As a result, the right operator realization preserves operator ordering, while the left realization reverses that order (cf. (2.13b,c)). Finally, the left and right realizations commute,

$$[\mathcal{D}^L(\hat{A}),\mathcal{D}^R(\hat{B})] = 0 \tag{2.15}$$

for arbitrary $\hat{A}$ and $\hat{B}$.

It is possible to transfer the action of an operator $\hat{A}$, in differential form, to the weighting function $F(\alpha)$ of an operator $\hat{F}$ using simple integration by parts

$$\hat{A}\hat{F} = \int \hat{A}|\alpha\rangle\langle\alpha|\, F(\alpha)\, \frac{d^2\alpha}{\pi} \tag{2.16}$$

$$= \int F(\alpha)\{\mathcal{D}^L(\hat{A})|\alpha\rangle\langle\alpha|\}\, \frac{d^2\alpha}{\pi} \tag{2.17}$$

$$= \int |\alpha\rangle\langle\alpha|\{\tilde{\mathcal{D}}^L(\hat{A})\, F(\alpha)\}\, \frac{d^2\alpha}{\pi} + \Gamma \quad . \tag{2.18}$$

In general, the surface term Γ vanishes. The adjoint realizations $\tilde{\mathcal{D}}$ are obtained from the realizations $\mathcal{D}$ by an algorithm expressed by the substitutions

$$\begin{aligned} \mathcal{D} &\to \tilde{\mathcal{D}} \\ \alpha &\to \alpha \\ \frac{\partial}{\partial\alpha} &\to -\frac{\partial}{\partial\alpha} \quad . \end{aligned} \tag{2.19}$$

In addition, operator ordering is reversed:

$$\mathcal{D}^L(\hat{B}\hat{A}) \to \tilde{\mathcal{D}}^L(\hat{A}\hat{B}) \quad . \tag{2.20}$$

Thus, the realization $\tilde{\mathcal{D}}^R$ obeys the properties (2.13), and in addition

$$\tilde{\mathcal{D}}^L(\hat{A}) = \left(\tilde{\mathcal{D}}^R(\hat{A}^\dagger)\right)^* \tag{2.21}$$

$$[\tilde{\mathcal{D}}^L(\hat{A}),\tilde{\mathcal{D}}^R(\hat{B})] = 0 \tag{2.22}$$

B. Extension to the Multimode $\mathcal{D}$-Algebra

The $\mathcal{D}$-operator algebra can be extended to the electromagnetic field with r independent modes. The creation and annihilation operators for such a multimode system obey the commutation relations

$$
\begin{aligned}
[a_j, a_k^\dagger] &= I\delta_{jk} \\
[a_j, a_k] &= 0 = [a_k^\dagger, a_j^\dagger] \\
[a_j, I] &= 0 = [I, a_j^\dagger]
\end{aligned}
\tag{2.23}
$$

Multimode coherent states for r independent modes are direct products of single mode coherent states

$$
\begin{aligned}
|\underset{\sim}{\alpha}\rangle &= |\alpha_1\rangle\, |\alpha_2\rangle \cdots |\alpha_r\rangle \\
&= e^{-\alpha^*\alpha/2} \sum_{m_1} \cdots \sum_{m_r} \frac{(\alpha_1)^{m_1} \ldots (\alpha_r)^{m_r}}{\sqrt{m_1! \ldots m_r}}\, |m_1, \ldots m_r\rangle \; .
\end{aligned}
\tag{2.24}
$$

Consequently, multimode projectors are direct products of single mode projectors

$$
|\underset{\sim}{\alpha}\rangle\langle\underset{\sim}{\alpha}| = \prod_{j=1}^{r} |\alpha_j\rangle\langle\alpha_j| \; . \tag{2.25}
$$

The construction of multimode $\mathcal{D}$-algebras follows easily from the corresponding single mode $\mathcal{D}$-algebras:

$$
\hat{A}\hat{F} = \int \hat{A}|\underset{\sim}{\alpha}\rangle\langle\underset{\sim}{\alpha}|\, F(\underset{\sim}{\alpha})\, \frac{d^2\underset{\sim}{\alpha}}{\pi^r} \tag{2.26}
$$

$$
= \int F(\underset{\sim}{\alpha})\, \{\mathcal{D}^L(\hat{A})\, |\underset{\sim}{\alpha}\rangle\langle\underset{\sim}{\alpha}|\}\, \frac{d^2\underset{\sim}{\alpha}}{\pi^r} \tag{2.27}
$$

$$
= \int |\underset{\sim}{\alpha}\rangle\langle\underset{\sim}{\alpha}|\, \{\mathcal{D}^L(\hat{A})\, F(\underset{\sim}{\alpha})\}\, \frac{d^2\underset{\sim}{\alpha}}{\pi^r} \tag{2.28}
$$

$$
\mathcal{D}^L(a_j^\dagger a_k) = \mathcal{D}^L(a_k)\, \mathcal{D}^L(a_j^\dagger) \tag{2.29}
$$

$$
= \alpha_k \left(\frac{\partial}{\partial\alpha_j} + \alpha_j^*\right) \tag{2.30}
$$

$$
\tilde{\mathcal{D}}^L(a_j^\dagger a_k) = \left(-\frac{\partial}{\partial\alpha_j} + \alpha_j^*\right)\alpha_k \; . \tag{2.31}
$$

The $\mathcal{B}$-algebra realizations $\mathcal{B}^L$ and $\tilde{\mathcal{B}}^L$ obey the properties (2.13). The left and right multimode algebras obey (2.14) and (2.15) while their adjoints obey (2.21) and (2.22).

C. Multilevel Systems

The $\mathcal{B}$-operator algebras have been used successfully in the c-number mapping of the master equation for a field mode coupled to a thermal reservoir,[3] an ensemble of N identical two-level atoms radiating into the vacuum of photons in one quasi-mode of the radiation field,[4] and the thermal relaxation of a 2J + 1-level system.[5] It is therefore useful to look for an analog algebra suitable for applications to general multilevel atomic or molecular systems of identical particles.

We consider the atomic or molecular system having r not necessarily equally spaced states, and call the operator describing transitions from state j to state k, E_{kj}. The complete set of transition operators forms an algebra which obeys commutation relations isomorphic to those of the Lie algebra u(r),

$$[E_{jk},E_{mn}] = E_{jn}\delta_{km} - E_{mk}\delta_{nj} \quad . \tag{2.32}$$

Coherent states for a system of N identical r-level systems are obtained by applying a unitary transformation to the total system ground state,

$$\left|{N \atop \underset{\sim}{x}}\right\rangle = U(\underset{\sim}{x})\left|{N \atop \text{Ground}}\right\rangle \quad . \tag{2.33}$$

In the special case of N identical 2-level systems, these are the now familiar ACGT atomic coherent states.[6]

$$\left|{N \atop x_1x_2}\right\rangle = \sum_{n_1+n_2=N}\left[\frac{N!}{n_1!n_2!}\right]^{1/2}(x_1)^{n_1}(x_2)^{n_2}\left|{J \atop m}\right\rangle \tag{2.34}$$

where

$$2J = n_1 + n_2$$

$$2m = n_1 - n_2$$

$$x_1^2 + |x_2|^2 = 1 \tag{2.35}$$

and the phase has been chosen such that x_1 is real.

The generalization of the ACGT atomic coherent states to multilevel coherent states is straightforward,[7]

$$|{}^{N}_{x}\rangle = \sum_{n_1+\ldots+n_r=N} \left[\frac{(n_1 + \ldots + n_r)!}{n_1!\ldots n_r!}\right]^{1/2} (x_1)^{n_1}\ldots(x_r)^{n_r} \times |n_1,\ldots n_r\rangle \tag{2.36}$$

with the constraint

$$x_1^2 + |x_2|^2 + \ldots + |x_r|^2 = 1 \quad . \tag{2.37}$$

There are analogous properties between the multilevel and multimode coherent states, and these properties have been discussed extensively elsewhere.[8] The most important property of the multilevel coherent states as far as our purposes here are concerned is their overcompleteness. Consequently, an arbitrary bounded operator can be expanded in terms of multilevel diagonal projectors,

$$\hat{F} = \int |{}^{N}_{x}\rangle\langle{}^{N}_{x}| \, F(x) \, dx \quad . \tag{2.38}$$

where the element of volume in this integral is given by

$$\underset{\sim}{dx} = \delta\left(\sum_{k=1}^{r} x_k^* x_k - 1\right) d^2x_1\ldots d^2x_r = d^2x_2\ldots d^2x_r \quad . \tag{2.39}$$

Because of the diagonal expansion property (2.38), it is useful to construct $\mathcal{D}$-operator algebras for multilevel systems in direct analogy with the $\mathcal{D}$-operator algebras for multimode systems, i.e.,

$$\hat{A}\hat{F} = \int \hat{A}|{}^{N}_{\underset{\sim}{x}}\rangle\langle{}^{N}_{\underset{\sim}{x}}|F(\underset{\sim}{x}) \, \underset{\sim}{dx} \tag{2.40}$$

$$= \int F(\underset{\sim}{x}) \left\{\mathcal{D}^{L}(\hat{A}) \, |{}^{N}_{\underset{\sim}{x}}\rangle\langle{}^{N}_{\underset{\sim}{x}}|\right\} \underset{\sim}{dx} \tag{2.41}$$

$$= \int |{}^{N}_{\underset{\sim}{x}}\rangle\langle{}^{N}_{\underset{\sim}{x}}|\left\{\tilde{\mathcal{D}}^{L}(\hat{A}) \, F(\underset{\sim}{x})\right\} \underset{\sim}{dx} + \Gamma \quad . \tag{2.42}$$

Here, as in the multimode case, the operator algebras for $\mathcal{D}^L$ and $\tilde{\mathcal{D}}^L$ are related by a simple integration by parts, and here as before the surface term Γ generally vanishes.

It is possible to construct the multilevel $\mathcal{O}$-algebras by direct computation. However, it is simpler and more useful to construct these algebras from the multimode $\mathcal{O}$-algebras using the homomorphism described in the next part of this section.

D. Multimode to Multilevel Homomorphism

A very useful homomorphism relates the full multimode algebra, consisting of all distinct operator products (up to second order) of the form 1, a_j, $a_k^\dagger$, $a_j^\dagger a_k$, a_j, a_k, $a_k^\dagger a_j^\dagger$ with the multilevel algebra (2.32). The algebra homomorphism h_A is given explicitly by

$$\begin{aligned} a_j a_k, a_k^\dagger a_j^\dagger &\xrightarrow{h_A} 0 \\ a_j^\dagger a_k &\longrightarrow E_{jk} \\ a_j, a_h^\dagger, 1 &\longrightarrow 0 \quad . \end{aligned} \tag{2.43}$$

The algebra homomorphism can be used to provide a coherent state homomorphism under the parameter identification

$$\begin{bmatrix} \alpha_r \\ \cdot \\ \cdot \\ \cdot \\ \alpha_2 \\ \alpha_1 \end{bmatrix} = \begin{bmatrix} x_r \\ \cdot \\ \cdot \\ \cdot \\ x_2 \\ x_1 \end{bmatrix} \alpha\, e^{i\phi} \tag{2.44}$$

where we have explicitly extracted the phase ϕ according to

$$\frac{\alpha_1}{|\alpha_1|} = e^{i\phi} \tag{2.45}$$

and α is defined as

$$\sum_{j=1}^{r} |\alpha_j|^2 = \alpha^2 \quad . \tag{2.46}$$

Under this change of variables, the multimode coherent states (2.24) can be reexpressed in the equivalent form

$$|\underset{\sim}{\alpha}\rangle = e^{-\alpha^{\dagger}\alpha/2} \sum_{m_1\ldots m_r} \frac{(\alpha_1)^{m_1}\ldots(\alpha_r)^{m_r}}{\sqrt{m_1!\ldots m_r!}} \left|m_1,\ldots m_r\right\rangle$$

$$= e^{-\alpha^2/2} \sum_{N=0}^{\infty} \frac{(\alpha\, e^{i\phi})^N}{\sqrt{N!}} \sum_{n_1+\ldots+n_r=N} \left[\frac{(n_1+\ldots+n_r)!}{n_1!\ldots n_r!}\right]^{1/2}$$

$$\times\, (x_1)^{n_1}\ldots(x_r)^{n_r}\left|n_1,\ldots n_r\right\rangle \quad . \qquad (2.47)$$

The homomorphism from the field coherent state $|\underset{\sim}{\alpha}\rangle$ to the atomic coherent state $|{}^{N}_{\underset{\sim}{x}}\rangle$ is given explicitly by

$$|\underset{\sim}{\alpha}\rangle \xrightarrow{h_c} |{}^{N}_{\underset{\sim}{x}}\rangle$$

$$h_c = \lim_{\alpha\to 0} \left\{\frac{\partial}{\partial(\alpha\, e^{i\phi})}\right\}^N \frac{e^{\alpha^2/2}}{\sqrt{N!}} \quad . \qquad (2.48)$$

The coherent state homomorphism h_c can be used to construct a projector homomorphism h_p. The r-mode projectors $|\underset{\sim}{\alpha}\rangle\langle\underset{\sim}{\alpha}|$ are functions of the r complex variables $\alpha_1,\alpha_2,\ldots,\alpha_r$, whereas the r-level projectors $|{}^{N}_{\underset{\sim}{x}}\rangle\langle{}^{N}_{\underset{\sim}{x}}|$ are functions of the r-1 complex parameters $x_2,\ldots,x_r$. The two extra real variables, α and ϕ, provide the freedom needed to construct h_p explicitly. If $|\underset{\sim}{\alpha}\rangle\langle\underset{\sim}{\alpha}|$ is integrated with respect to ϕ, the result connects only multilevel states with the same N value,

$$\oint |\underset{\sim}{\alpha}\rangle\langle\underset{\sim}{\alpha}| \frac{d\phi}{2\pi} = e^{-\alpha^2} \sum_{N=0}^{\infty} \frac{(\alpha^2)^N}{N!} |{}^{N}_{\underset{\sim}{x}}\rangle\langle{}^{N}_{\underset{\sim}{x}}| \quad . \qquad (2.49)$$

The projector homomorphism is therefore given explicitly by

$$|\underset{\sim}{\alpha}\rangle\langle\underset{\sim}{\alpha}| \xrightarrow{h_p} |{}^{N}_{\underset{\sim}{x}}\rangle\langle{}^{N}_{\underset{\sim}{x}}|$$

$$h_p = \lim_{\alpha^2\to 0} \left(\frac{\partial}{\partial\alpha^2}\right)^N e^{\alpha^2} \oint \frac{d\phi}{2\pi} \quad . \qquad (2.50)$$

The Lie algebra homomorphism (2.43) provides a $\mathcal{D}$-algebra homomorphism h_D as follows:

$$
\begin{aligned}
\mathcal{D}^L(E_{jk}) &= \mathcal{D}^L(a_j^\dagger a_k) && \text{by (2.43)} \\
&= \mathcal{D}^L(a_k)\ \mathcal{D}^L(a_j^\dagger) && \text{by (2.29)} \\
&= \alpha_k \left(\frac{\partial}{\partial \alpha_j} + \alpha_j^* \right) && \text{by (2.30)}
\end{aligned}
\tag{2.51}
$$

Under the change of variables (2.44), we easily obtain the $\mathcal{D}$-algebra homomorphism

$$
h_D = x_k \left\{ \frac{\partial}{\partial x_j} + x_j^* \left[\left(\alpha^2 + \frac{\partial}{\partial \alpha^2} \right) - \frac{1}{2} \left(\underset{\sim}{x} \cdot \underset{\sim}{\nabla} + \underset{\sim}{x}^* \cdot \underset{\sim}{\nabla}^* \right) \right] \right\}
\tag{2.52}
$$

where

$$
\underset{\sim}{x} \cdot \underset{\sim}{\nabla} + \underset{\sim}{x}^* \cdot \underset{\sim}{\nabla}^* = \sum_{k=2}^{r} \left(x_k \frac{\partial}{\partial x_k} + x_k^* \frac{\partial}{\partial x_k^*} \right) ,
\tag{2.53}
$$

and we have defined

$$
\frac{\partial}{\partial x_1} = \frac{1}{2x_1} \left(\frac{\partial}{\partial (i\phi)} - \underset{\sim}{x} \cdot \underset{\sim}{\nabla} + \underset{\sim}{x}^* \cdot \underset{\sim}{\nabla}^* \right) = - \frac{\partial}{\partial x_1^*}
\tag{2.54}
$$

consistent with the constraint (2.37) for the $r - 1$ independent complex variables x_k. The results and interrelations of these four homomorphisms are illustrated graphically:

$$
\begin{array}{ccccl}
\mathcal{D}^L(a_j^\dagger a_k) & |\underset{\sim}{\alpha}\rangle\langle\underset{\sim}{\alpha}| & = & \Big\{ \alpha_k & \left(\dfrac{\partial}{\partial \alpha_j} + \alpha_j^* \right) \Big\} |\underset{\sim}{\alpha}\rangle\langle\underset{\sim}{\alpha}| \\
h_D \big\downarrow \ \ h_A \big\downarrow & h_P \big\downarrow & & h_c \big\downarrow & \big\downarrow \text{chain rule} \qquad h_P \big\downarrow \\
\mathcal{D}^L(E_{jk}) & |{}^N_{\underset{\sim}{x}}\rangle\langle{}^N_{\underset{\sim}{x}}| & = & x_k \Big\{ \dfrac{\partial}{\partial x_j} & + x_j^* [N - \frac{1}{2} (\underset{\sim}{x} \cdot \underset{\sim}{\nabla} + \underset{\sim}{x}^* \cdot \underset{\sim}{\nabla}^*)] \Big\} |{}^N_{\underset{\sim}{x}}\rangle\langle{}^N_{\underset{\sim}{x}}|
\end{array}
\tag{2.55}
$$

Here, the operator $\partial/\partial x_1$ is defined as in (2.54), except that the $\partial/\partial(i\phi)$ term is absent due to the integration over ϕ in h_P, (c.s. (2.50)).

It is to be noted that since the multilevel $\mathcal{D}$-algebras are projected homomorphically from the multimode $\mathcal{D}$-algebras, the multilevel $\mathcal{D}$-algebras possess the same abstract properties (2.13), through (2.15), (2.19) through (2.22).

In the next section, we make application of the coherent states and $\mathcal{D}$-algebras for multilevel systems to a particular problem of current interest in quantum statistical mechanics. We shall use the results of this section to develop the mapping of operator equations of motion to equivalent c-number partial differential equations.

III. APPLICATION TO c-NUMBER MAPPING OF OPERATOR EQUATIONS OF MOTION

A. Dynamical Operator Equations of Motion

In this section, we present an application of the results of the previous section in connection with the solutions of two kinds of equations of motion which commonly arise in quantum statistical mechanics. For illustrative purposes, we shall specialize here to the particular case where the system operators are of the angular momentum type, i.e., ensembles of equivalent two-level atoms.[9]

The first case to be considered is a class of master equations of the form

$$\dot{\hat{W}} = \sum_{m,n} C_{mn} \hat{A}_m \hat{W} \hat{B}_n \tag{3.1}$$

where the C_{mn} are c-number coefficients and the $\hat{A}_m$ and $\hat{B}_n$ are combinations of angular momentum operators. The second is the equation of motion for the diagonal element Q of the reduced density operator in the atomic coherent state[6] $|\Omega\rangle$,

$$Q \equiv \langle\Omega|\hat{W}|\Omega\rangle \quad , \tag{3.2}$$

defined by

$$\frac{\partial}{\partial t} Q(t) = \frac{\partial}{\partial t} \mathrm{tr}[\hat{W}(t)\ \hat{\Lambda}(\Omega)] = \sum_{m,n} C_{mn}\ \mathrm{tr}[\hat{A}_m \hat{W}(t)\ \hat{B}_n\ \hat{\Lambda}(\Omega)], \tag{3.3}$$

where $\hat{\Lambda}(\Omega)$ is the atomic coherent state diagonal projector.

B. Mapping of the Master Equation

For r-level systems, the density operator $\hat{W}$ in (3.1) can always be written in its diagonal form in the coherent state basis.[4-6]

$$\hat{W}(t) = \int d\Omega \quad P(\Omega,t)\ \hat{\Lambda}(\Omega) \tag{3.4}$$

where $P(\Omega,t)$ is a quasi-probability c-number weighting function on the diagonal projector $\hat{\Lambda}(\Omega)$. We call the representation (3.4) the "P"-representation of the operator $\hat{W}$. If (3.4) is used in (3.1), the operators $\hat{A}_m$ and $\hat{B}_n$ on the right-hand side of the equation act on the projector $\hat{\Lambda}(\Omega)$.

This allows us to use the results of the previous section to replace the operators on the right-hand side of this equation by their c-number equivalents. Thus, using the homomorphism (2.55), eq. (3.1) can be written in the equivalent form

$$\int d\Omega \left\{ \hat{\Lambda}(\Omega) \frac{\partial P(\Omega,t)}{\partial t} - P(\Omega,t) \sum_{m,n} C_{mn} \mathcal{D}^L(\hat{A}_m) \mathcal{D}^R(\hat{B}_n) \times \hat{\Lambda}(\Omega) \right\} = 0 \quad . \tag{3.5}$$

The action of the $\mathcal{D}$-operators in this equation can be transferred to the c-number distribution function $P(\Omega,t)$ as indicated in section II.A by integrating by parts. If this is done, the surface terms, as already noted, in general give zero contribution and (3.5) becomes

$$\int d\Omega\, \hat{\Lambda}(\Omega) \left\{ \frac{\partial P(\Omega,t)}{\partial t} - \sum_{m,n} C_{mn} \tilde{\mathcal{D}}^L(\hat{A}_m) \tilde{\mathcal{D}}^R(\hat{B}_n) P(\Omega,t) \right\} = 0 \quad , \tag{3.6}$$

where we have used the identity

$$\int d\Omega\, P(\Omega,t) \mathcal{D}^L(J^\alpha) \hat{\Lambda}(\Omega) = \int d\Omega\, \hat{\Lambda}(\Omega) \tilde{\mathcal{D}}^L(J^\alpha) P(\Omega,t) \quad . \tag{3.7}$$

Here, the operators $\hat{A}_m$ and $\hat{B}_n$ are taken to be combinations of elementary angular momentum operators J^α. The algorithm relating $\mathcal{D}$ and $\tilde{\mathcal{D}}$' is obtained from the relations (2.19) and the homomorphism (2.55), and is given here explicitly as

$$\begin{aligned} &\mathcal{D}^L \longrightarrow \tilde{\mathcal{D}}^L \\ &N \longrightarrow N + r \\ &\frac{\partial}{\partial x_j} \longrightarrow -\frac{\partial}{\partial x_j} \qquad j = 1,2,\ldots,r \\ &\text{add } -\delta_{kj} \end{aligned} \tag{3.8}$$

and, in addition, operator ordering is preserved. It is a simple matter to show that if

$$\hat{A}_m = \hat{J}^{\alpha_1} \hat{J}^{\alpha_2} \ldots \hat{J}^{\alpha_m} \tag{3.9}$$

and

$$\hat{B}_n = \hat{J}^{\alpha_1}\hat{J}^{\alpha_2}\ldots\hat{J}^{\alpha_n} \quad (3.10)$$

then

$$\tilde{\mathcal{D}}^L(\hat{A}_m) = \tilde{\mathcal{D}}^L(\hat{J}^{\alpha_1})\ldots\tilde{\mathcal{D}}^L(\hat{J}^{\alpha_m}) \quad (3.11)$$

and

$$\tilde{\mathcal{D}}^R(\hat{B}_n) = \tilde{\mathcal{D}}^R(\hat{J}^{\alpha_n})\ldots\tilde{\mathcal{D}}^R(\hat{J}^{\alpha_1}) \quad . \quad (3.12)$$

It is to be noted that the order of the operators $\tilde{\mathcal{D}}^L$ and $\tilde{\mathcal{D}}^R$ in (3.6) is immaterial since they commute (c.s. (2.22)).

For the master eq. (3.1) to be satisfied, it is sufficient that the c-number part of the integrand appearing in (3.6) within the curly brackets be identically zero. This results in the c-number partial differential equation for the quasi-probability density distribution function $P(\Omega,t)$,

$$\frac{\partial P(\Omega,t)}{\partial t} = \sum_{m,n} C_{mn}\,\tilde{\mathcal{D}}^L(\hat{A}_m)\,\tilde{\mathcal{D}}^R(\hat{B}_n)\,P(\Omega,t) \quad . \quad (3.13)$$

The solution to this partial differential equation for the initial condition of a given system completely specifies the density operator (3.4) for all time. Expectation values for dynamical operators $\hat{X}_k$ of the atomic system are calculated using (3.4) and the relation,

$$\langle \hat{X}_k(t)\rangle = \mathrm{tr}[\hat{W}(t)\,\hat{X}_k]$$

$$= \int d\Omega \quad P(\Omega,t)\langle\Omega|\hat{X}_k|\Omega\rangle \quad . \quad (3.14)$$

The matrix element in (3.14) can be easily evaluated using the disentangling theorem of ACGT.[6]

C. Partial Differential Equation of Motion for Q

The $\mathcal{D}$-algebra can be applied directly to the explicit elimination of the operators in (3.3), with the result

$$\frac{\partial Q(\Omega,t)}{\partial t} = \sum_{m,n} C_{mn}\,\mathcal{D}^R(\hat{B}_n)\,\mathcal{D}^L(\hat{A}_m)\,Q(\Omega,t) \ . \quad (3.15)$$

Here, again, it is emphasized that the order of the operators $\mathcal{D}^R$ a[nd] $\mathcal{D}^L$ is immaterial, since they commute, (c.s. (2.15)). In this, th[e]

Q-representation, expectation values of dynamical operators $\hat{X}_k$ are calculated according to

$$\hat{X}_k(t) = \mathrm{tr}\left[\frac{(2J+1)}{4\pi}\int d\Omega\ \hat{W}(t)\ \hat{X}_k\ \hat{\Lambda}(\Omega)\right]$$

$$= \frac{2J+1}{4\pi}\int d\Omega\ \mathcal{D}^L(\hat{X}_k)\ Q(\Omega,t) \quad . \tag{3.16}$$

where we have used the overcompleteness of the coherent states $|\Omega\rangle$, inserted the resolution of the identity in the atomic subspace[6];

$$I = \frac{2J+1}{4\pi}\int d\Omega\ \hat{\Lambda}\ (\Omega) \tag{3.17}$$

into the trace.

D. Convolution Between P and Q

The partial differential equations (3.13) and (3.15) are formally equivalent. This is made clear by comparing (3.14) and (3.16), i.e., the solution to either differential equation gives the same statistical information.

The functions P and Q are explicitly related to one another by taking the matrix element of (3.4) in the coherent state $|\Omega'\rangle$, to give

$$Q(\Omega',t) = \int d\Omega\ \ P(\Omega,t)|\langle\Omega'|\Omega\rangle|^2 \quad . \tag{3.18}$$

The Q and P functions are thus related by a convolution as shown in Appendix B. It is emphasized that this equation together with (3.14) and (3.16) shows explicitly that all statistical information about the atomic system is contained in Q as well as in P as solutions, with the appropriate initial conditions, of (3.15) and (3.13), respectively.

IV. CONCLUSION

Using the homomorphism between r-mode and r-level coherent states, we have established the former as generating functions for the latter in (2.47). An analogous relation carries over to the diagonal projectors (2.49). We have used the homomorphism (2.43) between multimode boson creation and annihilation operators and the operators for multilevel systems which satisfy the appropriate Lie algebra (2.32), together with the r-mode to r-level homomorphism (2.48) and (2.50), to generate a homomorphism between r-mode and r-level $\mathcal{D}$-algebras (2.52). The results are summarized in (2.55).

These results were applied in section III to the mapping of the master equation for a system of identical two-level atoms (3.1) into an equivalent c-number partial differential equation in the P-representation (3.13). The $\mathscr{D}$-algebra was further used to generate partial differential equations of motion for the diagonal elements of the atomic density operator in the coherent state representation (3.15). We call this the Q-representation. Using the coherent states and the -algebra, we showed explicitly that the P and Q representations are statistically equivalent (c.s. (3.14), (3.16)), and are related by (3.18) in the form of a convolution integral.

The $\mathscr{D}$-algebra for two-level atomic systems has been very successfully applied with excellent results which appear in three previous publications. The first was an application to the mapping and analysis of the master equation in the P-representation describing the collective spontaneous relaxation of N identical two-level atoms radiating into the vacuum of photons in one quasi-mode of the radiation field.[4] The second uses the c-number mapping in the P-representation for the description and analysis of the thermal relaxation of a 2J + 1-level system.[5] The third application was made to the formal calculation of multitime correlation functions for systems described by operators of the angular momentum type.[10] In this case, the $\mathscr{D}$-algebra, together with the c-number mapping of the master equation in the P-representation, facilitates the reduction of multitime correlation functions to classical-like integrals where the Green's functions appearing in the integrands are shown to be explicitly evaluated from solutions to the c-number partial differential equation in the appropriate time regimes.

On the basis of the applications mentioned above, it appears that equations of the form (3.13) are nearly always of the "diffusion" type.[11] In these cases, the differential equations have, in general, a direct physical interpretation. Consequently, a large amount of qualitative information can be gleaned about the system before the equation is even solved. In the cases cited, it is to be emphasized that the essential difference between the c-number representations generated by our method and those done by others is that our resulting partial differential equations are of finite order in the derivatives, the order being determined by the maximum order of products of shift operators appearing in the master equation.

APPENDIX A: CHANGE OF VARIABLES

It is necessary to express the r complex variables α_u and the derivatives $\partial/\partial\alpha_u$ $(u = 1,2,\ldots,r)$ in terms of the r − 1 complex variables x_j and the derivatives $\partial/\partial x_j$ $(j = 2,3,\ldots,t)$ and the two real variables α, ϕ, and $\partial/\partial\alpha$, $\partial/\partial\phi$.[3] The relations between the variables themselves is straightforward

$$\alpha_j = x_j \, \alpha e^{i\phi}$$

$$\sum_{u=1} \alpha_u^* \alpha_u = \alpha^2$$

$$\frac{\alpha_1}{|\alpha_1|} = e^{i\phi} \quad . \tag{A.1}$$

The relationship between the derivatives is most easily determined using the chain rule. For example

$$\frac{\partial}{\partial\alpha_u} = \frac{\partial\alpha}{\partial\alpha_u}\frac{\partial}{\partial\alpha} + \frac{\partial\phi}{\partial\alpha_u}\frac{\partial}{\partial\phi} + \sum_{j=2}^{r}\left[\frac{\partial x_j}{\partial\alpha_u}\frac{\partial}{\partial x_j} + \frac{\partial x_j^*}{\partial\alpha_u}\frac{\partial}{\partial x_j^*}\right] \quad . \tag{A.2}$$

The partial derivatives in (A.2) and its complex conjugate are most conveniently determined implicitly according to the following example:

$$\alpha^2 = \sum_{n=1}^{r} \alpha_u^* \, \alpha_u$$

$$2\alpha \, d\alpha = \sum_{u=1}^{r} [\alpha_u d\alpha_u^* + \alpha_u^* d\alpha_u]$$

$$\frac{\partial\alpha}{\partial\alpha_u} = \frac{\alpha_u^*}{2\alpha} \quad , \quad \frac{\partial\alpha}{\partial\alpha_u^*} = \frac{\alpha_u}{2\alpha} \quad . \tag{A.3}$$

The other partial derivatives may be determined in a similar way. It is possible to establish the following identities:

$$0 = \frac{\alpha_1^*}{2\alpha^2} x_j + \frac{\alpha_1^*}{2\alpha_1^*\alpha_1} x_j + \frac{\partial x_j}{\partial\alpha_1}$$

$$\frac{\delta_{jk}}{\alpha e^{i\phi}} = \frac{\alpha_k^*}{2\alpha^2} x_j + 0 + \frac{\partial x_j}{\partial\alpha_k}$$

$$0 = \frac{\alpha_1}{2\alpha^2} x_j - \frac{\alpha_1}{2\alpha_1^*\alpha_1} + \frac{\partial x_j}{\partial \alpha_1^*}$$

$$0 = \frac{\alpha_k}{2\alpha^2} x_j + 0 + \frac{\partial x_j}{\partial \alpha_k^*} \quad . \tag{A.4}$$

A similar set of equations involving x_1^* is obtained from the set (A.4) by complex conjugation. As a result we obtain, after a little algebra

$$\frac{\partial}{\partial \alpha_1} = \frac{1}{2} x_1 e^{-i\phi} \frac{\partial}{\partial \alpha} + \frac{1}{2x_1 e^{i\phi}} \frac{\partial}{\partial (i\phi)} - \frac{x_1 e^{-i\phi}}{2\alpha} (\underset{\sim}{x} \cdot \underset{\sim}{\nabla} + \underset{\sim}{x}^* \cdot \underset{\sim}{\nabla}^*)$$

$$- \frac{1}{2x_1 e^{i\phi}} (x \cdot \nabla - x^* \cdot \nabla^*) \tag{A.5}$$

$$\frac{\partial}{\partial \alpha_j} = \frac{1}{2} x_j^* e^{-i\phi} \frac{\partial}{\partial \alpha} + \frac{1}{\alpha e^{i\phi}} \frac{\partial}{\partial x_j} - \frac{x_j e^{-i\phi}}{2\alpha} (\underset{\sim}{x} \cdot \underset{\sim}{\nabla} + \underset{\sim}{x}^* \cdot \underset{\sim}{\nabla}^*) \quad . \tag{A.6}$$

In these equations

$$\underset{\sim}{x} \cdot \underset{\sim}{\nabla} \pm \underset{\sim}{x}^* \cdot \underset{\sim}{\nabla}^* = \sum_{j=2}^{r} \left[x_j \frac{\partial}{\partial x_j} \pm x_j^* \frac{\partial}{\partial x_j^*} \right] . \tag{A.7}$$

Expressions for $\partial/\partial\alpha_u^*$ are obtained from (A.5) and (A.6) by complex conjugation.

APPENDIX B: COHERENT STATE CONVOLUTION THEOREM FOR P AND Q

A relation of the form (3.18) between the P and Q representations obtains for the field coherent state representation as well as the atomic coherent state representation. These relations are given respectively by

$$Q(\alpha) = \int d^2\beta \;\; P(\beta) |\langle \beta | \alpha \rangle|^2 \tag{B.1a}$$

and

$$Q(\Omega) = \int d_\mu \Omega \; P(\Omega,t) | \langle \Omega' | \Omega \rangle |^2 \quad . \tag{B.1b}$$

We note that[1] $|\langle \beta | \alpha \rangle|^2 = e^{-|\alpha-\beta|^2}$ and therefore (B.1a) is a convolution and may therefore be inverted by standard techniques. Let q(k) be the Fourier transform of Q(α), i.e., { F(Q) = q(k)}. The convolution transformation theorem gives

$$q(k) = p(k)\, e^{-|k|^2/4} \tag{B.2a}$$

where we have used $F(e^{-|\alpha|^2}) = e^{-|k|^2/4}$. As a result, P(β) is given explicitly by

$$P(\beta) = F^{-1}\left\{q(k)/e^{-|k|^2/4}\right\}. \tag{B.3a}$$

A similar procedure is used to invert (B.1b). Since

$$\langle\Omega'|\Omega\rangle = \sum_m \left(\Gamma^j_{m,-j}(\Omega')\right)^* \Gamma^j_{m,-j}(\Omega) = \Gamma^j_{-j,-j}(\Omega'^{-1}\Omega) \quad ,$$

eq. (B.1b) has the form of a convolution on the Bloch sphere SU(2)/U(1). Consequently, (B.1b) may also be regarded as a convolution on the group SU(2). If $\tilde{F}$ represents the Fourier transform on SU(2), then the convolution theorem is expressed in the form[12]

$$\tilde{F}[Q(\Omega)] = \tilde{F}[P(\Omega)] \cdot \tilde{F}[|\langle Id|\Omega\rangle|^2] \quad . \tag{B.2b}$$

As a result, P(Ω) is given formally by

$$P(\Omega) = \tilde{F}^{-1}[\tilde{F}(Q)/\tilde{F}(|\langle Id|\Omega\rangle|^2)] \quad . \tag{B.3b}$$

To make (B.3b) more tractable, it is expedient to resolve the functions Q(Ω), P(Ω) and $|\langle\Omega'|\Omega\rangle|^2$ in terms of a complete set of functions defined on SU(2). Thus, P(Ω) becomes

$$P(\Omega) = P(\Omega h) = \sum_{m',m''} \frac{d(J)}{V(G)} P^J_{m'm''} \Gamma^J_{m'm''}(\Omega h) \tag{B.4}$$

where d(J) = 2J + 1 is the dimensionality of the unitary irreducible representation Γ^J of SU(2), V(G) is the volume of SU(2), h∈U(1), Ω∈SU(2)/U(1) and Ωh represents an arbitrary group element in SU(2). The Fourier coefficients of P(Ω) are $P^J_{m'm''}$, i.e.,

$$P^J_{m'm''} = \int P(\Omega h) \left(\Gamma^J_{m'm''}(\Omega h)\right)^* d_\mu\Omega\, d_\mu(h) \tag{B.5}$$

The Fourier coefficients of the weighting function $|\langle\Omega'|\Omega\rangle|^2$ are given by

$$R^J_{m'm''} = \int \Gamma^j_{-j-j}(\Omega h) \left(\Gamma^j_{-j-j}(\Omega h)\right)^* \left(\Gamma^J_{m'm''}(\Omega h)\right)^* d_\mu\Omega \frac{d_\mu(h)}{V(H)} \quad . \tag{B.6}$$

This integral can be expressed in terms of Clebsch-Gordon coefficients

$$R^J_{m'm''} = \frac{V(G)/V(H)}{2j+1} \left\langle \begin{matrix} j & J \\ -j & m' \end{matrix} \middle| \begin{matrix} j \\ -j \end{matrix} \right\rangle \left\langle \begin{matrix} j \\ -j \end{matrix} \middle| \begin{matrix} j & J \\ -j & m'' \end{matrix} \right\rangle \quad . \tag{B.7}$$

We note that the only nonvanishing terms in this expression are those for which J = L (integer), $0 \leq L \leq 2j$, $m' = m'' = 0$. Then[14] using $V(G)/V(H) = V[SU(2)/U(1)] = 4\pi$,

$$R^L_{00} = \frac{4\pi}{2j+1} \left| \left\langle \begin{matrix} j \\ -j \end{matrix} \middle| \begin{matrix} j & L \\ -j & 0 \end{matrix} \right\rangle \right|^2$$

$$\left| \left\langle \begin{matrix} j \\ -j \end{matrix} \middle| \begin{matrix} j & L \\ -j & 0 \end{matrix} \right\rangle \right|^2 = \frac{(2j+1)!(2j)!}{2j+1+L)!(2j-L)!} \quad . \tag{B.8}$$

The explicit form of (B.2b) is then

$$Q^L_{m0} = P^L_{m0} \, R^L_{00} \quad .$$

As a result

$$P(\Omega h) = \sum_{L=0}^{2j} \sum_{m=-L}^{+L} \frac{d(L)}{V(G)} \frac{Q^L_{m0}}{R^L_{00}} \, \Gamma^L_{m0}(\Omega h) \quad . \tag{B.9}$$

Since $P(\Omega h) = P(\Omega)$ and $\Gamma^L_{m0}(\Omega h) = \Gamma^L_{m0}(\Omega)$, the argument Ωh in (B.9) can be replaced by the coset representative Ω . In addition, (B.9) can be expressed in terms of the more familiar spherical harmonics using[15]

$$Y^L_m(\Omega) = \sqrt{\frac{d(L)}{V(G)/V(H)}} \, \Gamma^L_{m0}(\Omega) \tag{B.10}$$

$$\begin{aligned} Q^L_{m0} &= \int Q(\Omega h) \left(\Gamma^L_{m0}(\Omega h) \right)^* d_\mu(\Omega h) \\ &= V(H) \int Q(\Omega) \left(\Gamma^L_{m0}(\Omega) \right)^* d_\mu(\Omega) \qquad\qquad (B.11) \\ &= V(H) \left[\frac{V(G)/V(H)}{d(L)} \right]^{1/2} \int Q(\Omega) \left(Y^L_m(\Omega) \right)^* d_\mu(\Omega) \quad . \end{aligned}$$

As a result

$$P(\Omega) = \int K(\Omega,\Omega') \, Q(\Omega') \, d_\mu(\Omega') \quad , \tag{B.12}$$

where

$$K(\Omega,\Omega') = \frac{2j+1}{4\pi} \sum_{L=0}^{2j} \frac{1}{\left| \left\langle \begin{matrix} j \\ -j \end{matrix} \middle| \begin{matrix} j & L \\ -j & 0 \end{matrix} \right\rangle \right|^2} \times \sum_{m=-L}^{+L} Y^L_m(\Omega) \left(Y^L_m(\Omega') \right)^* \quad .$$

Since the kernel $K(\Omega,\Omega')$ is a function of the group operation $\Omega'^{-1}\Omega$ only (by the addition theorem for special harmonics), the inverted relation (B.12) has the form of a convolution. In fact, in the field coherent state representation, (B.3a) can also be put into the form of a convolution.

Finally we exhibit explicitly the symmetry between the convolution in (B.1b) and (B.12) as follows:

$$\begin{aligned} Q(\Omega) &= \int K_{+}(\Omega,\Omega')\ P(\Omega')\ d_{\mu}(\Omega') \\ P(\Omega) &= \int K_{-}(\Omega,\Omega')\ Q(\Omega')\ d_{\mu}(\Omega') \\ K_{\pm}(\Omega,\Omega') &= \sum_{L=0}^{2j} (R^{L}_{00})^{\pm 1} \sum_{m=-L}^{+L} Y^{L}_{m}(\Omega) \left(Y^{L}_{m}(\Omega')\right)^{*} . \end{aligned} \tag{B.13}$$

This result generalizes immediately to the case for the multilevel coherent states (r >2).

References

(1) R. J. Glauber, Phys. Rev. 130, 2529 (1963); 131, 2766 (1963).

(2) K. E. Cahill, Phys. Rev. B138, 1566 (1965).

(3) W. H. Louisell, Proc. Int. Sch. of Physics, E. Fermi Course 42. Quantum Optics, Edited by R. J. Glauber, Academic Press, 1969; sec. 1-2; M. Sargent III, M. O. Scully, W. E. Lamb, Jr., Laser Physics, Addison-Wesley (1974), sec. 16.2.

(4) L.M. Narducci, C. A. Coulter, and C. M. Bowden, Phys. Rev. A9, 829 (1974).

(5) L.M. Narducci, C. M. Bowden, V. Bluemel, and G. P. Carrazana, Phys. Rev. A11, 280 (1975).

(6) F. T. Arecchi, E. Courtens, R. Gilmore, and H. Thomas, Phys. Rev. A6, 2211 (1973).

(7) R. Gilmore, Ann. Phys. (N.Y.) 74, 391 (1972).

(8) R. Gilmore, Rev. Mex. de Fisica 23, 143 (1974).

(9) R. H. Dicke, Phys. Rev. 93, 99 (1954).

(10) L. M. Narducci, C. M. Bowden, V. Bluemel, G. P. Carrazana, and R. A. Tuft, Phys. Rev. A., March 1975, in press.

(11) In the cases mentioned, the c-number equations are of the Fokker-Planck type, except that the diffusion coefficient associated with the second partial derivative with respect to ϕ is nonpositive definite (Ref. 6, Sec. II and Ref. 10, Appendix B). The nonpositive definiteness of the diffusion matrix which arises can be removed in some cases on physical grounds by initially integrating over the variable ϕ in the diagonal representation of the density operator in the coherent state representation. See Ref. 4.

(12) Ref. 7, pp. 452-455.

(13) Ref. 6, eq. D20, p. 2235.

(14) Ref. 6, eq. D22, p. 2235.

(15) Ref. 6, Appendix D.

NONZERO TEMPERATURE VARIATIONAL PRINCIPLE APPLIED TO LOW TEMPERATURE LIQUID SODIUM*

R. L. Coldwell, M. A. Pokrant, and A. A. Broyles

Department of Physics and Astronomy,
University of Florida
Gainesville, Florida 32611

Abstract: A previously derived nonzero temperature variational principle is applied to liquid sodium at 630°K to test the idea of treating fluids as collections of nuclei and electrons. The effects of symmetrizing the wave functions are incorporated using an extension of Lado's method. The remainder of the Slater sum is parameterized using two-body effective potentials containing the analytic large-r and small-r limits and four additional parameters. The minimum of the variational integral is found by evaluating the multidimensional integrals, using the biased-selection Monte Carlo method, for 48 sets of parameter values. The resulting nucleon-nucleon distribution is compared with experiment and the nucleon-electron distribution with the ground-state, single atom Hartree-Fock result.

I. INTRODUCTION

The authors have previously reported[1] the derivation of a nonzero temperature variational principle for the Slater sum (diagonal density matrix) which minimizes the difference between the energy obtained from the free energy and the expectation value of the Hamiltonian averaged over states. This variational principle was shown[1] to reduce to the usual ground state variational principle for the lowest wave function in the zero temperature limit. Here we report on some preliminary calculations demonstrating the feasibility of using this variational principle by applying it to liquid sodium at a temperature of 630°K (0.004 Ry) and a density of 2.43×10^{22} nuclei/cm^3 ($r_s = 4.0484a_0$).

---

*Work supported in part by the National Science Foundation.

The functional

$$J[U] = \left\langle -\frac{\partial U}{\partial \beta} + (Y-Y_I) + V + \frac{1}{4}\ \Delta U\cdot\Delta U \right\rangle \tag{1}$$

is minimized with respect to arbitrary variations of U with $\partial U/\partial\beta$ held fixed.[1] We have defined

$$\Delta A\cdot\Delta B = \sum_i \frac{\hbar^2}{2m_i}\ \nabla_i A\cdot\nabla_i B \tag{2}$$

$$\langle A\rangle = \int e^{-U-S} A\, d^{3N}r \Big/ \int e^{-U-S} d^{3N}r \quad , \tag{3}$$

where the integration is over the coordinates of all of the particles. The quantity U+S is the negative logarithm of the Slater sum for the system

$$U+S = -\ \ell n[(\Pi_\nu N_\nu!\lambda_\nu^{3N}\nu)\sum_n (e^{-\beta H/2}\Psi_n)^*(e^{-\beta H/2}\Psi_n)] \ , \tag{4}$$

where the product is over all the types of particles present, $\lambda_\nu^2 = 2\pi\hbar^2\beta/m_\nu$, and $\beta = 1/kT$. The function S alone is the negative logarithm of the Slater sum for the corresponding ideal gas. This term S, which contains the effects of the symmetrization of the wave functions, is an input function from the solvable ideal gas system and is not determined here.

The term Y comes from the fact that we have not been able to form a differential equation in temperature for the Slater sum entirely in terms of itself.[1] The remaining terms had their zero and infinite temperature limits subtracted, leaving Y such that

$$Y-Y_I \xrightarrow[T\to 0]{} 0 \quad , \quad Y-Y_I \xrightarrow[T\to\infty]{} 0 \quad . \tag{5}$$

If we add the energy of the non-interacting system, E_I, to both sides of Eq. (1), we have

$$J[U]+E_I = \left\langle -\frac{\partial U}{\partial\beta}\right\rangle + \langle V\rangle + \left\langle \frac{1}{4}\Delta U\cdot\Delta U + E_I + Y - Y_I\right\rangle \tag{6}$$

The last term can be shown to be exactly the kinetic energy in the zero temperature limit and will be negligibly different for the temperatures considered here. This method then differs from the usual ground-state variational principle only through the term $\langle -\partial U/\partial\beta\rangle$.

The quantity $(Y-Y_I)$ could be computed by an independent method; e.g., a modified Thomas-Fermi method was used in Ref. 2. At the temperature considered here, T/T_F is extremely small for the electrons so that for them $Y-Y_I$ should be very close to its zero temperature limit, while the massive nuclei are essentially classical so that for them $Y-Y_I$ should be very close to its infinite temperature limit. For these reasons we will assume $Y-Y_I = 0$ here, so that Y disappears from the variational principle.

We assume U has the following parameterization:

$$U = \sum_{i<j} u_{nn}(r_{ij}) + \sum_{i,j} u_{ne}(r_{ij}) + \sum_{i<j} u_{ee}(r_{ij}) , \qquad (7)$$

where the sums are over nucleon-nucleon, nucleon-electron, and electron-electron pairs, respectively, and

$$u_{nn} = 4z^2/\tau R + 2\beta\varepsilon_o z^2 e^{-\alpha R}/R \qquad (8)$$

$$u_{ne} = -(4z/\tau R)(1-e^{-b_1 R}) - A(1-e^{-R^2/4})^2/R^4 \qquad (9)$$

$$u_{ee} = (4/\tau R)(1-e^{-b_2 R}) + B(1-e^{-\gamma R^2})/R^4 \qquad (10)$$

where $R=r/a_o$, $b_1^2=\tau$, $b_2^2=\tau/2$, $\tau=\Gamma\coth(\beta\Gamma\varepsilon_o/2)$, $\Gamma^2 = 16\pi\rho_e a_o^3$, and ε_o=1Ry. At the temperature and density considered here, $\tau \cong \Gamma = 1.415$. The first term in each of the three u's is suggested by the small-r and large-r parts which can be determined analytically.[2,3] The parameters α, A, B, and γ were varied until a minimum in J was found. The symmetry effective potential, S, was taken to be

$$S = \sum_{i<j} s^{(2)}(r_{ij}) = \sum_{i<j<k} s^{(3)}(r_{ij}, r_{ik}, r_{jk}) \qquad (11)$$

with these sums over parallel-spin electrons only. For $s^{(2)}(r)$ we used a fit to Lado's approximation.[4,5] For $s^{(3)}$ we used[6]

$$s^{(3)}(r_{12}, r_{13}, r_{23}) = -\ln[g_I^{(3)}(r_{12}, r_{13}, r_{23})/g_I(r_{12})g_I(r_{13})g_I(r_{23})]. \qquad (12)$$

It seemed necessary to include $s^{(3)}$ to account for the tremendous local density variation in going from the K-shell to the interatomic region.

At the temperature considered here (630°K), it is reasonable to assume that the electron-electron and nucleon-electron u's have quit changing with temperature since the atoms are completely formed and the temperature is far less than the Fermi temperature of the electrons. On the other hand, the temperature is high enough so that the nucleon-nucleon u is essentially classical at small distances. Hence,

$$\frac{\partial U}{\partial \beta} \cong \sum_{i<j} \frac{\partial u_{nn}(r_{ij})}{\partial \beta} \cong \sum_{i<j} \frac{z^2 e^2}{r_{ij}} e^{-\alpha r_{ij}} , \qquad (13)$$

where α is the parameter to be determined variationally and self-consistently. That is, we choose $\alpha=\alpha_o$ in the term $\partial U/\partial\beta$ and hold this fixed. J is now minimized with respect to α, holding α_o fixed. To be self-consistent we must find $\alpha_{min}=\alpha_o$. The other three parameters need only be determined variationally since they do not enter $\partial U/\partial\beta$ at the temperature considered here.

Increasing the value of α decreases the repulsion between nucle at the average particle separation causing them to come closer together. Since g_{nn} is large at points near the repulsive barrier, increasing α will raise $\langle V \rangle$. The expectation value of the term $-z^2e^2\sum_{i<j}\exp(-\alpha_o r_{ij})/r_{ij}$, when evaluated over $U(\alpha)$, will lower $\langle -\partial U/\partial\beta \rangle$ as α increases. If the assumed α_o is too small, this lowering occurs faster than $\langle V \rangle$ increases for $\alpha>\alpha_o$. Thus the minimum of J will occur for some α greater than α_o, meaning it is not a self-consistent minimum. It can be seen that for some value of α_o, these tendencies will cancel, giving the minimum α equal to α_o, as desired for a self-consistent solution. It remains to demonstrate that α is independent of temperature, as assumed here. This can only be done by determining α in a similar manner at another temperature.

II. NUMERICAL METHODS AND RESULTS

The four-parameter variation was carried out at T = 630°K using the biased-selection Monte Carlo method.[7] We generated 6000 statistically independent configurations consisting of the positions of 32 nuclei and 352 electrons in a box periodically repeated. These configurations were stored along with the absolute probability, P, that they would have been generated by a random selection procedure. These probabilities varied over a range of e^{1000}. Each evaluation of the multidimensional integral in Eq. (1) then amounted only to determining U+S for each configuration and adding to it $\ell n P$ for that configuration, $\omega=\ell nP+U+S$. The 32 configurations with the largest value of ω would generally have a range of ω of about 70 out of a range of ω for the 6000 configurations on the order of 10,000. These top 32 configurations were treated as equal for purposes of determining J and the correlation functions. The 6000 configurations cost 20 seconds of running time apiece on an IBM 360/165, while the value of J could then be evaluated over the 6000 stored

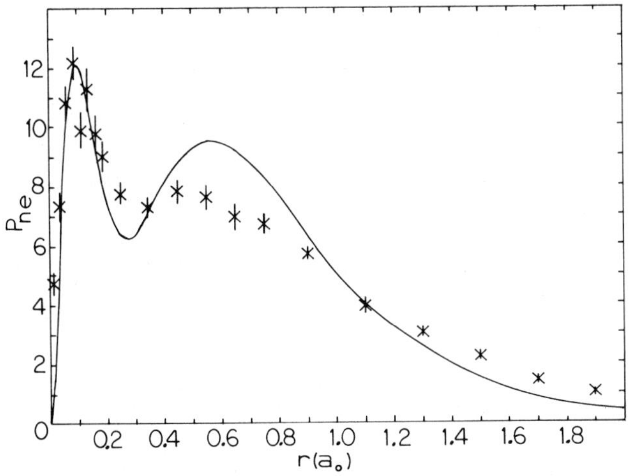

Figure 1. The radial density of electrons about the Sodium nucleus, $P_{ne}=4\pi\rho_e r^2 g_{ne}$, for $r_s=4.0484a_o$. The values of the variational parameters are $\alpha=1.0$, $\gamma=2.25$, A=64, and B=0.5. Solid line, Hartree-Fock result for a single atom in the ground state;[8] x's, present calculation, T=630°K.

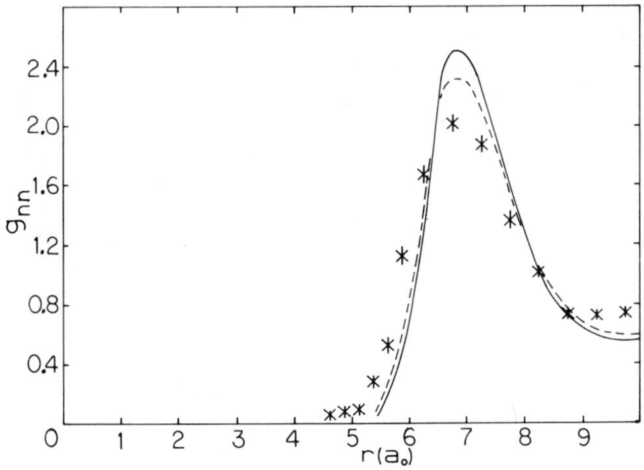

Figure 2. The nucleon-nucleon pair correlation function, g_{nn}, for liquid Sodium with $r_s=4.0484a_o$. The values of the variational parameters are $\alpha=1.0$, $\gamma=2.25$, A=64, B=0.5. Solid line, "experimental" x-ray scattering result[9] at T=373°K; dashed line, "experimental" x-ray scattering result[9] at T=473°K; x's, present calculation T=630°K.

configurations for about three minutes of running time.

The parameters were varied over 48 different sets of values in the search for a minimum in the four-dimensional parameter space. The data were smoothed by fitting to a four-dimensional polynomial with the sum of the coefficients squared constrained to be as small as possible under the restriction that the χ^2 be equal to the number of data points.

The correlation functions corresponding to the lowest smoothed J are given in Figs. 1 and 2. The energy found, using Eq. (33) of Ref. 1, under the assumption $Y=Y_I$ was $-(348\pm2)$Ry/atom. This is below the -323.23 Ry of Hartree-Fock calculations for the ground state of a single atom by far more than the correlation energy could amount to. The kinetic energy was found to be about 434Ry/atom while the potential energy was about -782Ry/atom, so that the virial theorem is not satisfied. Since our set of Monte Carlo configurations contained only a few that give g_{nn}'s more sharply peaked than those in Fig. 2, it is possible that our g_{nn} might not be as sharply peaked as it should be. The generation of more configurations would clarify this point.

III. CONCLUSIONS

These calculations indicate the feasibility of using the non-zero temperature variational principle to find the Slater sum at a finite temperature. The results also demonstrate the feasibility of treating fluids as collections of nuclei and electrons. The low energy calculated suggests that a better statistical potential is needed. It is possible to improve S in several ways. One of these[6] would be to include $s^{(4)}$ and $s^{(5)}$. More astute choices for the parameterization of U could also improve the results. In particular, we feel that if u_{ee} had a different shape, the L-shell in g_{ne} would agree more closely with the Hartree-Fock results. Still another possibility would be in include a three-body potential, u_{nee}. While the pressure cannot be computed directly from the virial theorem (which is not satisfied here), the particles are reasonably close to where they should be, and it is the authors' feeling that a set of locations approximating that which an instantaneous photograph of the system would give provides a basis for powerful approximation methods for computing thermodynamic quantities. Higher temperatures are currently being investigated.

References

(1) M. A. Pokrant, A. A. Broyles, and R. L. Coldwell, Int. J. Quantum Chem. 8S, 403 (1974).

(2) M. A. Pokrant, A. A. Broyles, and T. Dunn, Phys. Rev. A 10, 379 (1974).

(3) T. Dunn and A. A. Broyles, Phys. Rev. 157, 156 (1967); M. A. Pokrant and F. A. Stevens, Jr., Phys. Rev. A 7, 1630 (1973); F. A. Stevens, Jr. and M. A. Pokrant, Phys. Rev. A 8, 990 (1973); H. J. Shin, doctoral dissertation, University of Florida, 1973 (unpublished).

(4) F. Lado, J. Chem. Phys. 47, 5369 (1967).

(5) F. A. Stevens, Jr. and M. A. Pokrant, J. Chem. Phys. 61, 3768 (1974).

(6) M. A. Pokrant, J. Chem. Phys. (to be published).

(7) R. L. Coldwell, Phys. Rev. A 7, 270 (1973); R. L. Coldwell, T. P. Henry, and C. W. Woo, Phys. Rev. A 10, 897 (1974).

(8) Calculated from the wave functions given in the supplement to the paper by E. Clementi, IBM Journal of Research and Development 9, 2 (1965).

(9) The "experimental" g(r) is from a calculation by R. D. Murphy and M. L. Klein, Phys. Rev. A 8, 2640 (1973), whose structure factors compared well with the experimental structure factors of A. J. Greenfield, J. Wellendorf, and N. Wiser, Phys. Rev. A 4, 1607 (1971).

LIST OF CONTRIBUTORS

Abraham, B. M., Argonne National Laboratory

Adams, E. D., University of Florida

Ahlers, G., Bell Laboratories

Ambegaokar, V., Cornell University

Baym, G., University of Illinois

Bergman, D. J., Bell Laboratories and Tel-Aviv,University (permanent address)

Bowden, C. M., Redstone Arsenal, Alabama

Bozler, H. M., Cornell University

Britton, C. V., University of Florida

Broyles, A. A., University of Florida

Buchler, J. R., University of Florida

Coldwell, R. L., University of Florida

Coleman, A. J., Queen's University, Kingston, Ontario, Canada

Combescot, M., Cornell University

Combescot, R., Cornell University

Coon, S. A., Université de Liège, Belgium

Dahm, A. J., Case Western Reserve University

Dynes, R. C., Bell Laboratories

Edwards, D. O., The Ohio State University

Emery, V. J., Brookhaven National Laboratory

Fatouros, P. P., The Ohio State University

Fetter, A. L. Stanford University

Gilmore, R., University of South Florida

Greenberg, A. S., University of Florida and Colorado State University (permanent address)

Hallock, R. B., Brown University and University of Massachusetts (permanent address)

Ihas, G. G., The Ohio State University

Ketterson, J. B., Northwestern University and Argonne National Laboratory

Kirk, W. P., University of Florida

Kobe, D. H., North Texas State University

Kojima, H., University of California at San Diego

Kummer, R. B., University of Florida

Langenberg, D. N., University of Pennsylvania

Lawson, D. T., Duke University (permanent address) and Cornell University

Lee, D. M., University of Florida and Cornell University (permanent address)

Maki, K., University of Southern California

Mittag, L., University of Miami

Mrozinski, P., The Ohio State University and du Pont Experimental Station, Wilmington, Delaware (permanent address)

Mueller, K. H., Institut für Festkörperforschung, Kernforschungsanlage Jülich, W. Germany

Mueller, R. M., University of Florida

Narayanamurti, V., Bell Laboratories

Narducci, L. M., Worcester Polytechnic Institute

Paulson, D. N., University of California at San Diego

Pobell, F., Institut für Festkörperforschung, Kernforschungsanlage, Jülich, W. Germany

Pokrant, M. A., University of Florida

Rainer, D., Cornell University

Reinstein, L. E., Boston University and Fracis Bitter National Laboratory, M. I. T.

Roach, P. D., Argonne National Laboratory

Roach, P. R., Argonne National Laboratory

Saslow, W. M., Texas A & M University

Serene, J. W., Stanford University

Shen, S. V., The Ohio State University and Northwestern University (permanent address)

Stephen, M. J., Rutgers University

Tam, C. P., The Ohio State University

Wheatley, J. C., University of California at San Diego

Wölfle, P., Max-Planck-Institut für Physik und Astrophysik, München, W. Germany

Woo, C.-W., Northwestern University

Zimmerman, G. O., Boston University and Francis Bitter Natio Laboratory, M. I. T.

SUBJECT INDEX